Geographical Information Systems
for Urban and Regional Planning

The GeoJournal Library

Volume 17

Geographical Information Systems for Urban and Regional Planning

edited by

HENK J. SCHOLTEN

*Department of Economics, Free University,
Amsterdam, The Netherlands; &
National Institute for Public Health and Environmental Protection,
Bilthoven, The Netherlands*

and

JOHN C. H. STILLWELL

*School of Geography, University of Leeds,
Leeds, United Kingdom*

KLUWER ACADEMIC PUBLISHERS
DORDRECHT / BOSTON / LONDON

Library of Congress Cataloging-in-Publication Data

Geographical information systems for urban and regional planning /
   edited by Henk J. Scholten and John C.H. Stillwell.
        p.   cm. -- (GeoJournal library ; v. 17)
   Based on a summer institute held in Aug. 1989 at the Academie van
Bouwkunst.
   ISBN 0792307933
   1. Regional planning--Data processing--Congresses. 2. City
planning--Data processing--Congresses. 3. Land use--Data
processing--Congresses. 4. Information storage and retrieval
systems--Geography--Congresses.   I. Scholten, H. J. (Henk J.)
II. Stillwell, John.   III. Series.
HT391.G42   1990
307.1'2'0285--dc20   90-43651

ISBN 0-7923-0793-3

For our children

Preface

In August 1989, a Summer Institute was held at the Academie van Bouwkunst, the seventeenth century home of Amsterdam's School of Architecture, Town Planning and Landscape. The meeting brought together experts in Geographical Information Systems from throughout the world to address an international audience of planners. The contents of this book reflect many of the themes that were presented and discussed at the conference.

The Summer Institute, let alone this volume, would not have been possible without the support of the International Association for the Development and Management of Existing and New Towns (INTA/AIVN), the International Society of City and Regional Planners (ISoCaRP), The National Physical Planning Agency of the Netherlands (RPD) and the Berlage Studio. We wish to acknowledge the assistance provided by these organisations and by the various sponsors: The Ministry of Housing, Physical Planning and Environment, the Municipality of Amsterdam, Logisterion b.v., ESRI, UNISYS, MABON b.v., SPSS, PRIME Computer Inc., PANDATA. The provision of hardware facilities by the various computer companies allowed immensely valuable 'hands on' experience to be gained by all the participants.

We are indebted to colleagues from the National Physical Planning Agency in the Hague and from GEODAN in Amsterdam for their various contributions in preparation for and during the course of the Summer Institute. In particular, we wish to thank Michel Grothe for his work in preparing camera-ready copy of the manuscript, and we want to express our gratitude to Annemieke de Waart for all her efforts in organizing the conference and in facilitating the production of the book.

Henk Scholten and John Stillwell
May, 1990

CONTENTS

List of figures

List of tables

List of plates

EDITORIAL INTRODUCTION

John C.H. Stillwell and Henk J. Scholten

In his opening address to the 1989 Summer Institute on 'Information in Urban and Regional Planning' in Amsterdam, Yap Hong Seng, Director of the Berlage Institute, drew attention to the rapid development in technology that has characterised the transition into the current 'age of information'. Modern computers now provide a capability for integrating and analysing different types of geographically referenced data and of displaying this information in an effective and attractive fashion.

Urban and regional planning, by its very nature as an information-rich discipline, has much to gain from the facilities which the new technology provides. In different countries, the activities which comprise procedural and substantive planning have evolved over time, yet the basic ingredients of formulating hypotheses based on problem analysis, of identifying goals and objectives, of designing alternative strategies and formulating new policies, of implementing plans and programmes, and of monitoring and updating, remain intact. The effectiveness with which any one of these process activities is undertaken depends upon the availability and appropriate utilisation of information of different types which are obtained from various sources. Information organised within Geographical Information Systems (GIS) therefore provides a framework to support the processes of decision making which together constitute planning practise.

The information requirements of national, regional and local planning authorities vary extensively. The provision of local education or social services, for example, requires knowledge of the demographic and socioeconomic structure of the population of small areas, whereas the strategic planning of national or regional transportation infrastructure depends not only on information about topogaphy and existing land use, but on future forecasts of traffic volumes. Consequently, information needs and data availability vary significantly. In this context, one of the key benefits of GIS is in creating the environment within which data collected from different sources for the same set of areas or individuals can be related in such a way that new, previously unknown information can be generated. This function of adding value through data integration is one of the key benefits to be obtained from GIS and can be achieved in various ways through overlay techniques with locational information or through combining attribute information using statistical or modelling techniques.

Despite the fairly widespread recognition of the benefits of GIS in urban and regional planning, the adoption of GIS by planning authorities has not been particularly encouraging. In Europe, for example, very few local planning authorities have accepted the challenge to embrace the technology. One of the reasons for this reaction is apprehension due to the perceived complexity of much of the proprietary software and the associated lack of in-house expertise with which to make use of the systems now available. Whilst education within planning is necessary to train individuals with fundamental computing skills, considerable progress has been achieved in the development of interface systems that allow managers and executives without advanced technical skills to use GIS. The development of more user-friendly software is a trend that is occurring in the private sector as well as the public sector. Other factors with a negative influence on the decision to adopt GIS include the high cost of GIS as perceived by executive planners faced with increasing demand for limited resources and the problems of selecting which GIS is most appropriate for a particular application. There is a tendency to try and evaluate what the different systems currently available can provide, rather than to begin with the needs and practical requirements that are to be satisfied.

Rapid advances have taken place in GIS technology in recent years and these changes are ongoing. This book seeks to demonstrate how GIS is contributing to urban and regional planning at different spatial scales and in different sectoral contexts. It draws together descriptions, assessments and opinions of GIS development and application by professional planners, academic researchers and hardware and software vendors. The book is divided into eight parts containing twenty one chapters prepared by different authors.

In Part I, the fundamental features of a GIS are elaborated, its relationship with urban and regional planning is outlined and the forces which are generating its phenomenal growth worldwide are explained. Database management is the theme of Part II. A description of two microcomputer based information systems developed for use by planning authorities in the San Francisco area is followed by a comprehensive outline of the design of databases and developments in GIS at the Dutch National Physical Planning Agency. Part III contains three sections which focus on the application of GIS in urban planning. Firstly, the role of information provision and management within the process of planning is discussed in connection with the redevelopment plan for Amsterdam's harbour front, the IJ-axis. Secondly, the advantages and problems associated with the implementation of a GIS by a local authority are reviewed on the basis of the experience of the municipality of Amersfoort in the Netherlands. Thirdly, a detailed review is presented of the approach to using GIS to support the planning process by the Planning Department of the City of Tacoma in Washington State.

Both Parts IV and V are concerned with the development of GIS as decision support systems. The former contains four chapters which identify and outline the application of GIS techniques to assist problems of land use and environmental planning. Applications of ARC/INFO software are described which involve, firstly, the identification and evaluation of new sites for house construction in the Randstad, and secondly, the use of this software to assist in various environmental impact assessment studies in West Germany. Part IV also contains a discussion of the development of a system to support the planning of environmental zones around sources of pollution and an explanation of the use of multicriteria methods to study decision problems in the context of conflicting criteria. The latter contains an empirical application of multicriteria decision analysis to agricultural land use policies in the Netherlands. In Part V, the environmental focus is less pronounced and the emphasis switches to spatial analysis and modelling. This section of the book begins with a description of the use of GIS to analyse spatial patterns of crime in the City of Tacoma. The second chapter examines the need for a greatly enhanced degree of spatial analysis functionality in GIS, reviews current progress and defines a series of generic spatial analytical tasks for subsequent development and incorporation within GIS. The arguments in the third chapter of Part V suggest that most proprietary GIS packages are deficient in terms of their analytical and value adding capabilities, and that greater integration between GIS and modelling software is required since public and private sector organisations now demand more customized, user-friendly systems to carry out their strategic planning activities more effectively. The final chapter of Part V demonstrates how interactive simulation and optimization models can be integrated with new expert systems techniques of artificial intelligence, dynamic computer graphics and GIS to provide very sophisticated decision support systems.

Associated with the emerging awareness of the importance of geo-information, there is also a rapid worldwide growth taking place in the demand for GIS education. Although an increasing number of educational institutions offer regular GIS courses at different levels, there appears to be a general lack of well documented curricula and supporting educational material. The development of GIS curricula provides the subject matter for the first chapter in Part VI. The second chapter contains guidance and advice on the problems relating to data, personnel and institutional arrangements that are involved when implementing and managing a GIS system within an organisation.

Part VII contains two chapters which deal with developments in the hardware and software of GIS. In the first, the evolution of technology and of computerized spatial analysis from Computer Aided Design through to geoprocessing is outlined and trends in GIS hardware, data acquisition and standards are discussed. The second chapter deals more specifically with visualization and the development of methods for displaying output at various stages of geoprocessing. In the final chapter (Part VIII), a broader perspective is presented which summarises the changes taking place in this age of information and which presents the results of cross-national comparisons of planning and information systems.

We hope that this book will serve to establish that GIS is not just a passing fashion that preoccupies a small group of enthusiasts but a new and rapidly expanding discipline whose potential application in urban and regional planning is only just beginning to be realised.

PART I

GEOGRAPHICAL INFORMATION SYSTEMS AND PLANNING

1   GEOGRAPHICAL INFORMATION SYSTEMS: THE EMERGING REQUIREMENTS

Henk J. Scholten and John C.H. Stillwell

1.1   Introduction

Planning involves a variety of activities undertaken at different spatial scales by national, regional or local organisations. The preparation of aggregate national physical or economic plans contrasts with the processes of development control and land use allocation in small areas, whilst regional planning occurs at intermediate levels in the hierarchy. In the context of physical planning, conflicts arise because of the increasing demands for space from competing interests which typically include agriculture, housing, manufacturing and service industry, commercial property, public infrastructure and recreation. As towns and cities have continued to grow and expand, so the pressures on land have intensified, particularly at the fringes of the built-up area where conflict arises between the need to develop land and the need to preserve the natural environment. Although the price mechanism operates to determine land use to a certain extent, there exist in most societies a variety of control mechanisms which enable public authorities to allocate land in what is considered to be the most optimal way, so that everyone's interests are looked after as far as possible.

One of the important questions arising in physical planning concerns the type of instruments that can be utilised to assist the achievement of land use optimality in a situation when demands for housing, employment, recreation and so on are changing significantly. New technological developments lead to new opportunities. Technological change alters the concept of accessibility, for example, and new product and process innovations influence the production of goods and the provision of services. Measures to improve health and the quality of life are now receiving greater attention along with the need to protect and improve the environment. New techniques must therefore enable public authorities to gain insights into the consequences of decisions relating to investment in housing construction, industrial and infrastructure development and environmental management. It is also necessary to provide the capacity to review the current situation in which new developments are proposed.

In order to achieve these objectives, information is vital for effective guidance in this rapidly changing context. Planning therefore is considered to be an information processing activity. All relevant information must be stored, managed, made available and presented in a suitable form for use at different stages in the planning process. The information may be of various types, extensive in quantity, variable in quality and referring to areal units of different size. Geographical Information Systems (GIS) provide the frameworks within which these activities can be undertaken. The example of land use planning has been used to illustrate the rationale for GIS, but there are inevitably a whole range of alternative contexts in which applications of GIS can be beneficial to organisations in the private and the public sectors. Large retailing companies, for example, require strategies for opening new shops on the basis of information on potential purchasing power and store accessibility, while public companies providing electricity must plan to install supply cables on the basis of information which relates to ease of maintenance and extension.

H. J. Scholten and J. C. H. Stillwell (eds.),
Geographical Information Systems for Urban and Regional Planning, 3–14.
© 1990 *Kluwer Academic Publishers. Printed in the Netherlands.*

At the same time that processes of industrial production have become automated, so have attempts been made to automate the handling of spatial information. Significant advances have occurred during the last decade. In this chapter, we seek to establish what this automation means. Hardware and software developments are briefly reviewed in Section 1.4, types of spatial information system are outlined in Section 1.5 and the variety of application fields, organisations and users are discussed in Section 1.6. We begin, however, by clarifying the objectives of GIS and outlining the basic types of data and data storage.

1.2   The objectives of GIS

Geographical information systems represent a technology designed to achieve particular objectives. In recent years, a variety of GIS products to assist the management and manipulation of spatial and non-spatial data have arrived on the market and users worldwide have begun to gain familiarity with these new systems. Experience suggests that there can be no doubt that the application of GIS is making significant contributions in facilitating the availability, integration and presentation of information.

Burrough (1983) has provided one of the most widely quoted definitions of GIS as "a powerful set of tools for collecting, storing, retrieving at will, transforming and displaying spatial data from the real world". As a technology, a GIS is not necessarily limited to the confines of one independent system. It may well have several components, each with a particular objective. We can therefore identify three main tasks which a GIS must accomplish. Firstly, there is the storage, management and integration of large amounts of spatially referenced data. A spatially referenced database can be conceived as containing two types of information, locational and attribute data (Figure 1.1). Locational or spatial data are two or three dimensional co-ordinates of points (nodes), lines (segments) or areas (polygons). Non-locational or descriptive data, on the other hand, refer to the features or attributes of points, lines or areas. Data is obtained from a wide variety of sources and one of the most important features of GIS is the facility to integrate data, converting data values to a common spatial framework, for example.

The second main objective of GIS is to provide the means to carry out analyses which relate specifically to the geographic component of the data. The analysis techniques may be simple or more sophisticated. At the simplest level, for example, data about different spatial entities such as soil type (per square kilometre) and land use (by local administrative area) can be combined by overlay analysis. At an intermediate level, GIS may allow statistical calculations of the relationship between data sets to be computed or distances between entities may be used to determine the route that must be followed to move as quickly as possible from one location to another. The most sophisticated analysis occurs when modelling is introduced. In this context, there are a variety of analytical opportunities. It is possible, for example, to use atmospheric modelling techniques to discover which areas might be affected by pollution resulting from an explosion at a particular hazardous installation (e.g. Chernobyl), given certain wind and weather conditions. Alternatively, modelling methods can be used to determine the impact of opening a new superstore or of locating a large public facility (e.g. a hospital) at different sites in a city region.

The third main task of GIS involves the organisation and management of large quantities of data in such a way that the information is easily accessible to all users. A GIS must also be able to display data on maps of high quality. Maps no longer have to be drawn by hand; they are an implicit product of all the work that is carried out within a GIS. However, for many different purposes, other forms of display (e.g. graphs and tables) may also be required, often for use in combination with maps. The RIA (Ruimtelijke Informatie via Automatisering) system, developed by the National Physical Planning Agency in the Netherlands (Scholten and Meijer 1988) is an excellent example of alternative methods of displaying planning information in an interactive and user-friendly way.

Figure 1.1: Locational and attribute information

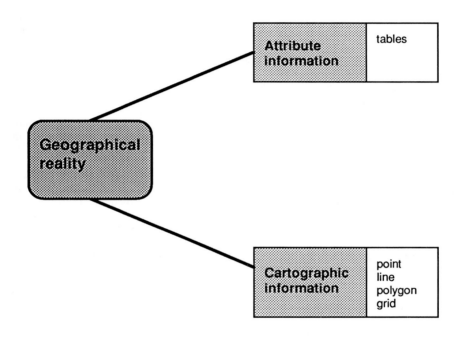

1.3 Types of data storage

The distinction has already been drawn between locational and attribute data. It is important to distinquish further between three forms in which locational data can be incorporated within a GIS: raster, vector and quadtree storage (Figure 1.2).

Raster or grid storage

This form of storage for locational data involves a regular grid of cells being laid over an area. Attribute data are collected for each grid cell which may measure 500 by 500 square metres, for example. Thus, the area is covered by a group of cells each of which has an attribute value. Within a grid-oriented system of this type, it is often the case that only limited use is made of the attribute data. Satellite photographs, in which a considerably smaller grid size is used, provide raster information. In a satellite photograph, a single value is attached to each cell. In this way, factual data can be collected in a very efficient way. Curran (1985) indicates how remote sensing data is frequently contained in raster based GIS.

Vector storage

The storage of locational data in vectors gives a very precise representation of reality. In this way, points, lines and areas are incorporated as Cartesian co-ordinates in the computer. Whilst lines can only be represented as a series of cells in a raster structure, in the vector storage method, the exact middle point of a line can be identified. An area is represented by a group of cells with an angular boundary in the raster structure, whereas the precise boundary of an area can be included when a vector structure is used. This precise storage of x and y co-ordinates usually generates larger data sets. The relationship between locational and attribute data is of great importance within a vector system. Each element is related to a record in the database with the same, unique identification number.

Quadtree storage

The quadtree storage method falls between the grid and vector storage methods. In this method, the data are stored in grid cells of variable size. Within larger homogeneous areas, a large cell size is used whereas towards the edges of the area, the cell size diminishes to form a precise picture. An area is therefore covered by considerably fewer quadtree cells of varying size than is the case in the regular grid storage method. The speed with which analyses can be carried out with a quadtree structure is high, whilst the original precision of the data is retained rather well.

Figure 1.2: The three forms of data storage

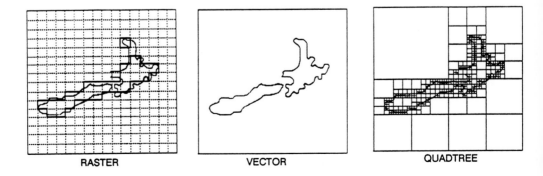

RASTER   VECTOR   QUADTREE

1.4   Hardware and software in a GIS environment

The initial GIS products date from the 1970s, a decade in which the typical hardware configuration comprised a central computer surrounded by memory and storage disks and a number of peripheral devices. Time share systems enabled a large number of terminals with lines attached to the mainframe to be used at the same time. At the beginning of the 1980s, this centralised approach was extended by connecting minicomputers to the central mainframe in order to carry out certain processes. This was the period in which GIS came of age. Very large databases began to be assembled and the need for processor capacity increased enormously. In the middle of the decade, the personal computer (PC) arrived, although its impact for GIS meant little more than an extra terminal in many cases. Nevertheless, it goes without saying that the PC has become central to the popularisation of GIS.

In the second half of the 1980s, the PC played an important role in simple GIS tasks such as the automatic production of maps. It is exactly this function that has brought GIS to the attention of many people, whilst the basic concept of a common central database tends to be forgotten. Attempts to transfer fully-fledged GIS from the mainframe onto the PC have been commercially successful, even though in many cases their performance leaves something to be desired. Of much greater importance in the mid-1980s was the further development of minicomputers and work stations connected to a network. At this time, the larger organisations that were making use of GIS had realised that the central mainframe option was not the solution to a number of GIS tasks (Figure 1.3). It was recognized that each separate task required its own processor capacity or its own working environment. Throughout the 1980s, it became clear that hardware vendors were meeting these demands perfectly well by means of further optimization of minicomputers using servers and the increased processor capacity of work stations, and that hardware costs were falling significantly.

Identification of the main tasks of a GIS environment allows us to specify a corresponding set of software demands.

Database management software

One of the most significant advances in GIS software development came at the end of the 1970s with the introduction of the concept of the relational database. In the relational model, different data sets are linked together by the use of common key fields. For example, attribute data available for two different sets of spatial units (areas) will require a third set of information to show how the two spatial bases fit together. This type of structure can also be used to construct spatial databases in which lines are linked together to represent polygons.

The creation, maintenance and accessing of a database requires a Data Base Management System (DBMS). In order to handle very large quantities of information, a relational DBMS is a necessity for most GIS applications (see Frank 1988). It also serves the very important function of separating the data user from the technicalities of the computing system and facilitates data manipulation and analysis. Many of the proprietary GIS have their own systems to handle basic storage, management and analysis operations (e.g. INFO is the relational DBMS of ARC/INFO; ORACLE is the relational DBMS of ARGIS).

Analysis software

The software required to perform certain analytical tasks varies according to the nature of the problem, the quantity of information available and the objectives of the organisations involved.

Figure 1.3: Decentralized GIS tasks

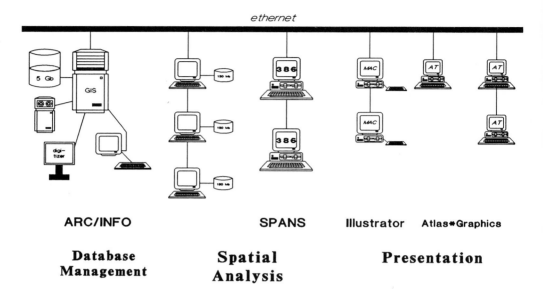

A variety of analytical tools are now available within GIS. The overlay procedure, for example, has been widely used for combining different data sets in order to identify areas or sites with required characteristics (see, for example, Dangermond 1983). Buffering, address matching and network analysis are additional tools adopted in planning applications of GIS. However, the development of analytical functionality within GIS has tended to neglect the important benefits that modelling procedures can contribute through data transformation, integration and updating; simulation; optimization, impact assessment; and forecasting (see Birkin et al. 1987). The specialized and complex nature of modelling algorithms and the restrictions imposed by processor capacity have both been influential in keeping the modelling component separate. One of the challenges confronting the next generation of GIS is to improve the integration of modelling and GIS so as to provide decision-makers and planners with enhanced model based decision support systems.

Access and presentation software

One of the most important functions of GIS development is frequently the provision of access to information by a variety of different users. In the context of planning, individuals in different departments of the same organisation (Public Works, Social Services, Transport, Parks, etc.) may require access to the same database. Similarly, users in different national, regional and local organisations may wish to access the database simultaneously. The National On-line Manpower Information System (NOMIS) in the United Kingdom is a good example of a GIS which stores up-to-date information about employment, unemployment, job vacancies, migration and population, and enables users at remote sites to obtain raw counts, tables, graphs and maps of data at a variety of spatial scales (Townsend et al. 1986, 1987).

The technological advances in computer hardware which have been occurring over the past 20 years have had a direct impact on the presentation of information in GIS. There have been striking advances in automated mapping, the development of programs to enable maps to be produced automatically (see Croswell and Clark 1988). The experience gained over this period has meant that professional products are now the norm (Plate 1.1), and micro-based mapping software (ATLAS*Graphics, GIMMS, MICROMAP, for example) provide high quality output.

1.5 Types of information systems

A good deal of discussion has taken place concerning the definition of a geographic information system. The GIS literature contains many terms that are used as synonyms for GIS which include spatial information system; geo-data system; geographic data system; land use information system (Clarke 1986, Parker 1988). This plethora of meanings arises in part because GIS is a very young science and has inportant connections with other physical and social science disciplines that involve handling spatial data. These include remote sensing, photogrammetry, cartography, surveying, geodesy, environmental science, regional science, planning and, of course, geography. Burrough's definition of GIS, quoted earlier in the chapter, is therefore one of a number of definitions that appear in the literature (see also, for example, Marble et al. 1983, Calkins and Tomlinson 1984, Berry 1986). In this discussion, for convenience, we use a rather rigid classification framework to illustrate the important differences between spatial information systems which fall under the GIS umbrella. Figure 1.4 gives an overview of the three main groups of spatial information systems that may be distinguished.

Computer Aided Design (CAD) systems

CAD systems are graphics systems which are used by industrial designers, architects and landscape architects to support and display their work. CAD has replaced the drawing board. Whilst early CAD systems were purely automated drawing systems, later packages have provided enhanced facilities for qualitative and quantitative design analysis as well as database facilities in which information can be stored and in which a large number of symbols can be used. CAD systems allow the possibility of automated drawing, the manipulation of drawings (changes of scale, location, zooming in, rotating and editing) and the presentation of this information in a professional format. Graphics software development has been under considerable influence from the world of CAD, and GIS presentation software has tried to incorporate CAD features. The developments in GIS have tended to lag behind those taking place in the CAD environment. It is important to recognize that the basic concepts in the two worlds are different. GIS is totally involved with the concept of the database whereas CAD is more concerned with the design process and the accompanying use of symbols. Both the hardware and software in the CAD environment are focused on presentation, whereas this principle applies less within the framework of GIS.

Automated Mapping (AM) was developed initially as another application of computer graphics technology alongside CAD. However, AM was extended to permit the storage and retrieval of associated locational and attribute data linked to the graphics, thus creating Automated Mapping/Facilities Management (AM/FM) for use as a specialist application in the management of utilities.

Figure 1.4: Spatial information systems

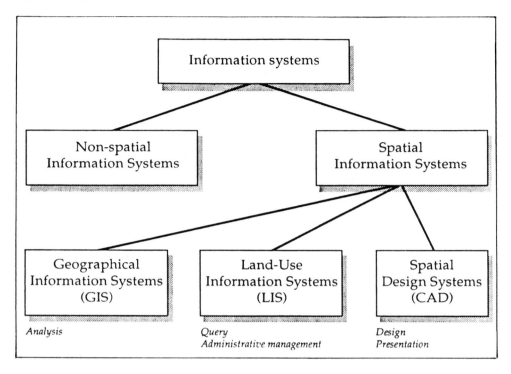

Land use Information Systems (LIS)

The objective of this type of information system is to function as an administrative system for the management of geographic data on land use and in this sense, LIS is in several ways similar to AM/FM. Many different demands are often made of LIS which have a direct influence on the way in which data should be stored. In the example of real estate information, concern over the legal obligations related to the precision of data may be of paramount importance. However, with regard to information on pipelines or cables, for instance, there may be less worry about legal matters but much more concern that the exact location of a pipe or cable can be established quickly. In this type of information system, the central focus is the development of a very detailed database. Moreover, there are tools available in this type of system which allow data with a very high degree of accuracy (double precision) to be stored, managed, integrated, updated and displayed. Relatively few spatial or geographical analyses are carried out within this type of system.

Geographical Information Systems (GIS)

Whilst LIS provide powerful tools for local planning authorities and public infrastructure agencies operating at a very detailed, micro scale, GIS tend to support analysis, planning and evaluation at a more macro scale. They are used in a variety of contexts to assist the research required to formulate and evaluate central and local government policy with respect to different aspects of physical and environmental planning on the one hand, and economic or strategic planning on the other. Such information systems adopt both a database approach and a set of tools to assist data collection, management, updating and presentation. One difference between GIS and LIS is in the order of magnitude of precision of the data that is held in the system, and the operations undertaken with the data. Another distinguishing feature of GIS in comparison with LIS and CAD is the availablity in GIS of the spatial analytical tools that have been discussed in a previous section.

These three categories of spatial information system are certainly not mutually exclusive as far as data are concerned. Many public utilities and local authorities have, for example, begun to use CAD systems to illustrate their information because CAD systems fulfill the demands for precision. However, the problem that confronts this group of users is the addition of information and its relationship to associated attribute information. A local authority or municipality may input its inventory of street lamps into a CAD system which possesses a number of qualities which are indispensible in the design process, for example. However, it is clear that this information may be related to other land use information and thus the integrated use of the information requires a database approach for which a LIS or GIS is required.

1.6   Applications and users of GIS

Despite efforts to distinguish different forms of spatial information system, it is commonplace to use the term GIS to refer to all types of spatial information systems. Applications of GIS are many and varied (see Department of Environment 1987). The following lists exemplify some of the fields of application of GIS which may help to confirm the distinguishing features of categories of information system that have been identified.

(i)   CAD and automated cartography
- civil engineering
- construction
- architecture
- landscape architecture
(ii)   LIS and facility management
- public utilities (water, gas, electricity, telephone)
- real estate (land and property)
- management of infrastructure (roads, railways, water supply, etc), housing, listed buildings, industry
(iii) . GIS
- traffic and transport planning
- agricultural planning
- environment and natural resource management
- recreation planning
- location/allocation decisions
- spatial planning (land use)
- service planning (education, social services, police, etc.)
- marketing

A number of categories of user can be identified on the basis of the objectives of the organisation to which the user belongs. A number of types of organisation can be classified which differ from one another according to the type of activity that each performs.

Four main types of organisation may be distinguished. Firstly, there is the research institute where research may be carried out to find solutions to problems or answers to questions posed by external paymasters. Data collection and manipulation takes place and descriptive, explanatory and predictive analyses are undertaken. Secondly, there are administrative institutions such as public utilities or property registration agencies. Here, the objective is to manage information in such a way that the process of acquiring and manipulating data is made as simple as possible. The management of waste disposal system of sewerage pipes for a local administrative area is one example where accurate and quick answers are required on the basis of the information stored in the GIS to questions such as: 'Where are the oldest parts of the system?'; 'What is the total length of pipe involved?'; 'At what depth are the pipes buried?'; and 'How many houses are connected to these pipes?'.

The third type of organisation is the government agency whose objective is to formulate policy recommendations. For this purpose, concept design and evaluation takes place. Commercial enterprises are the fourth type of organisation. Their aim is to maximize their profits by selling goods and services. Information is collected and manipulated within an integrated modelling/GIS environment to establish optimum locations for new retail outlets, for example.

The type of GIS that is adopted and applied therefore varies between each category of organisation and between organisations of different size and function within the same category. However, it is possible to identify particular groups of individuals across the spectrum of organisations whose occupational characteristics with respect to GIS are distinctive: information specialists, researchers, research coordinators, policy preparers, decision makers and third parties.

In each of the four categories of organisation, information specialists are required to acquire and manage data, computer hardware and software. The information specialist usually works with the raw data and requires a large GIS (e.g. ARC/INFO, ARGIS, SYSTEM9) which is flexible and able to be connected to other systems. Researchers, on the other hand, tend to be confined to their own institutes or to commercial companies. They work either with raw data or data that has been partially processed or transformed. They demand user friendliness from a GIS, analytical features (as with SPANS, for example) and appropriate interfaces that allow transfer of information to other packages for modelling and other purposes. Research coordinators are concerned with the interrelationship between the different products of the organisation and therefore work with manipulated data and require a simple, user friendly GIS. Policy formulation is usually the responsibility of a government agency and the main requirement of a GIS in this context is that it should be easy to use. The same applies to the decision makers in administrative, government and commercial organisations, whose job it is to translate information into policy statements, and to third parties, who simply utilise information provided by government agencies or research institutes.

The successful implementation of a GIS therefore depends on a careful preliminary assessment of the type of GIS required to meet the demands to be satisfied by the various users of the system. Whilst GIS packages are now available for the range of GIS users that we have identified, there remain considerable difficulties, particularly in large organisations, in establishing the configuration of hardware and software and in enabling the easy exchange of information between the machines and packages involved. It is important to ensure that there is a GIS management framework which takes into account the separation of functions between user departments and the central data store. Figure 1.5 shows the different user groups and their varying demands.

Figure 1.5: User groups and their demands

	information demand	user demand	type of GIS	development
infospecialist	raw data	analysis/ flexibility	large "open"	links to other packages
researcher	raw data/ pre-treated data	analysis accessibility	compact manageable	macro languages interfaces to other packages
management/ decisionmaker	(strategic) information	accessibility weighing / optimalisation	small and beautiful	user friendly interfaces
target group/ others	information	accessibility	small and beautiful	user friendly interfaces

1.7 Conclusion

In this first chapter, we have introduced a number of the emerging requirements of GIS that will be discussed in greater detail and exemplified more fully in the chapters which follow. The evidence contained in the rest of the book is testament to the fact that GIS is playing a key role in a variety of urban and regional planning activities across the world. It is a dynamic technology with enormous potential for the future. However, this potential will only be realised if those in the planning profession in senior administrative and executive positions are prepared to meet the challenges which GIS adoption entails and are able to demonstrate the vision necessary to create a suitable environment for successful GIS implementation.

References

Berry, J.K. (1986) Learning computer-assisted map analysis, Journal of Forestry, October, 39-43

Birkin, M., Clarke, G.P., Clarke, M. and Wilson, A.G. (1987) Geographical information systems and model-based locational analysis: ships in the night or the beginnings of a relationship?, Working Paper 498, School of Geography, University of Leeds

Burrough, P.A. (1986) Principles of Geographical Information Systems for Land Resources Assessment, Monographs on Soil and Resources Survey 12, Clarendon Press

Calkins, H.W. and Tomlinson, R.F. (1984) Basic Readings in Geographic Information Systems, SPAD Systems Ltd, Williamsville, New York

Clarke, K.C. (1986) Recent trends in geographic information system research, Geo-Processing, 3, 1-5

Crosswell, P.L. and Clark, S.R. (1988) Trends in automated mapping and geographic information system hardware, Photogrammetric Engineering and Remote Sensing, 54(11), 1571-1576

Curran, P.J. (1985) Principles of Remote Sensing, Longman

Dangermond, J. (1983) Selecting new town sites in the United States using regional databases, in Teicholz, E. and Berry, B.J.L. (eds) Computer Graphics and Environmental Planning, Prentice Hall Inc., Englewood Cliffs

Department of Environment (1987) Handling Geographic Information, HMSO, London

Frank, A.V. (1988) Requirements for a database management system for a GIS, Photogrammetric Engineering and Remote Sensing, 54(11), 1557-1564

Marble, D.F. and Peuquet, D.J. (1983) Geographic information systems and remote sensing, Manual of Remote sensing, 2nd Edition, American Society for Photogrammetry and Remote Sensing, Falls Church, Virginia

Parker, H.D. (1988) The unique qualities of a geographic information system: a commentary, Photogrammetric Engineering and Remote Sensing, 54(11), 1547-1549

Scholten, H.J. and Meijer, E. (1988) From GIS to RIA: a user-friendly microcomputer-orientated regional information system for bridging the gap between researcher and user, in: Polydorides, N., URSA-NET Proceedings 1988, Athens

Townsend, A., Blakemore, M., Nelson, R. and Dodds, P. (1986) The National On-line Manpower Information System (NOMIS), Employment Gazette, 94, 60-64

Townsend, A., Blakemore, M., Nelson, R. and Dodds, P. (1987) The NOMIS database: availability and uses for geographers, Area, 19, 43-50

Henk J. Scholten
Vrije Universiteit Amsterdam
Faculteit der Economische Wetenschappen en Econometrie
Postbus 7161
1007 MC Amsterdam
The Netherlands

John C.H. Stillwell
University of Leeds
School of Geography
Leeds LS2 9JT
United Kingdom

2 THE APPLICATION OF GEOGRAPHICAL INFORMATION SYSTEMS IN URBAN AND REGIONAL PLANNING

Henk F.L. Ottens

2.1 Introduction

This chapter addresses the possibilities for and the problems of bringing GIS technology into action for spatial planning purposes. Both urban and regional planning and geographical information systems are however changing and evolving rapidly. Some introductory remarks on the present-day nature of the two elements that make up the theme of this contribution are therefore necessary.

In practice, operational geographical information systems that already form a vital and well integrated part of a planning machine hardly exist. GIS technology holds many promises for urban and regional planners, but the introduction of GIS also reveals a lot of pitfalls. This intricate relationship between planning and GIS will be commented on, followed by a structured overview of possibilities for the use of geographical information systems in the different phases and for the different activities of the planning process. Relevant developments that can be expected in the near future are briefly reviewed at the end of the chapter.

2.2 The nature of urban and regional planning

In this discussion, we will restrict ourselves to planning activities that are initiated by government agencies. The private sector is however of importance. Large real estate development, financing and management firms do perform planning activities themselves and are often deeply involved in public-private partnership projects. But in most countries of the world, public agencies have been assigned prime responsibility for urban and regional planning.

The term 'planning' covers a wide range of activities. In the planning literature, strategic planning is often distinguished from operational planning (Kreukels 1980). Strategic planning has to do with long and medium term decision making. It often involves a lot of research, discussions, consultations and negotiations. The activities that support decision making can be split into two groups: the organization of the decision making process itself and the production of tangible results, in the form of plans, programmes and project initiatives. These two aspects of strategic planning are often referred to as procedural and substantive planning (Van der Cammen 1980). Procedural planning produces the organizational and decision making infrastructure in which substantive planners can produce research reports, policy reports, information material and, eventually, official plans, plan revisions, implementation programmes and project outlines. Initiatives and control activities connected with the implementation of plans that are in force are called operational planning or action-oriented planning (Wissink 1986). This planning aspect involves the judgement of applications and the issuing of permits with respect to development, building and installation. But it can also include the monitoring and control of projects in progress. At higher tiers of government, the evaluation and approval of planning activities performed by lower governmental bodies, is part of operational planning.

15

H. J. Scholten and J. C. H. Stillwell (eds.),
Geographical Information Systems for Urban and Regional Planning, 15–22.
© 1990 *Kluwer Academic Publishers. Printed in the Netherlands.*

Although spatial planning can be subdivided along the lines indicated in Figure 2.1, in practice, planning activities do interlink considerably. Further, it should be stressed that urban and regional planning normally is also integrated into a larger planning framework. Especially planning activities with respect to infrastructure, housing, retailing, public services and the environment are closely related to urban and regional planning.

Figure 2.1: Environment and aspects of spatial planning

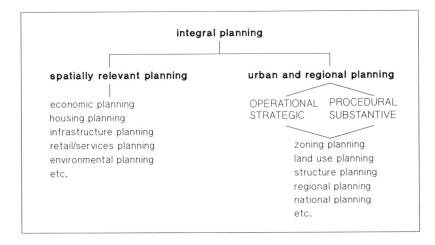

2.3 Planning cultures

The way in which planning activities are performed is not static over time. Each time period is dominated by a kind of 'paradigm', which emphasizes specific themes and procedures. These 'planning cultures' are also important for the methods and techniques that are most often employed. During the post-war period, three forms of planning can be distinguished (Kreukels 1985). Figure 2.2 gives an indication of the differences between these planning cultures.

Blueprint planning is highly technical in nature. This form of planning dominated the scene during the 1950s and 1960s. Planning activities were almost exclusively concentrated in building departments of state, provincial or municipal governments. Large scale urban extension schemes and major infrastructure works were the dominant type of projects realized during this period. In the United States, already during the mid-sixties this technical planning culture was gradually succeeded by a more social type of planning. The involvement of social scientists became much larger and special planning departments were established. Small scale, but comprehensive projects in old and new residential neighbourhoods attracted most attention and citizen participation became a major element of the planning process. In Europe this type of planning remained during the first half of the present decade. In North America a new planning culture, strategic planning, emerged during the late 1970s, with the important aim of promoting and supporting urban revival. Typically, a selective approach, concentrating on a few, but large and economically important projects, is chosen. Public-private cooperation, attention to urban and regional identity and emphasis on architectural beauty are other characteristics. In the last five to ten years, most West European countries have also adopted the strategic planning strategy.

Figure 2.2: Planning cultures

blueprint planning   technical, bureaucratic, technical infrastructure

process planning   social, democratic, neighbourhoods, urban renewal

strategic planning   selective, projects, public–private, urban revival

2.4   The state of the art of GIS

Geographical information systems have reached a mature stage. Based on fundamental research at universities, a lot of practical pioneering in government agencies and thanks to considerable investments by private software firms, off-the-shelf packages can now be bought that perform well in a production environment. Hard- and software costs have come down to a level such that a widespread acceptance of the GIS technology is imminent. GIS has become an umbrella concept for all automated systems that integrate and handle geo-referenced information. But this tendency can not only be observed in the terminology, it also is present in the design of operational systems (Dangermond 1988). Gradually, modular built multi-purpose packages, that can be configured or customized for specific applications, have come to dominate the market. In Figure 2.3, an overview is given of GIS modules that can be found in most full fledged systems.

The basis module of a geographical information system performs the general housekeeping of geographical information. Provisions for inputting data, producing simple map output, storing and managing graphical and alphanumeric data are always a part of the basis system. A modern GIS has interfaces with general purpose database packages like ORACLE, INGRES or INFORMIX and provides a toolbox with import, export and conversion programmes. To the basis module, a number of other modules can be added. These modules are designed to perform tasks in specific application areas. The Automated Mapping/Facility Management (AM/FM) systems form, for example, a well established branch of GIS. AM/FM systems, which are used for repetitive tasks and operational management, have become more and more integrated with other business software systems, like document processing and accounting. Analytical GIS systems often have special modules for spatial analysis, geostatistics and spatial modelling and can process both raster and vector data. There are new developments with respect to dynamic cartography and vision systems. Finally the design features of GIS are greatly extended and image processing systems are integrated with GIS. A special and still separate position is held by the positioning and navigation systems, a kind of real time GIS, used for routing and traffic control.

So far, little progress has been made in the fields of dedicated decision support and expert systems based on the GIS technology. Advances in these fields would greatly improve the functionality of GIS for direct policy making activities. Also project management software has not yet been integrated with GIS. By combining GIS modules, several types of operational geographical information systems can be configured. When those systems are classified by the themes on which information is processed, a family of geographical information systems emerges. Figure 2.4 gives a possible typology of such a GIS family (Braedt 1988). Environmental Information Systems (EIS), Land Information Systems (LIS) and Urban and Regional Information Systems (URIS) form the major entries into this scheme. Within the group of urban and regional information systems, a distinction is made between systems which handle technical information and those dealing with socio-economic data.

Figure 2.3: The family of GIS modules

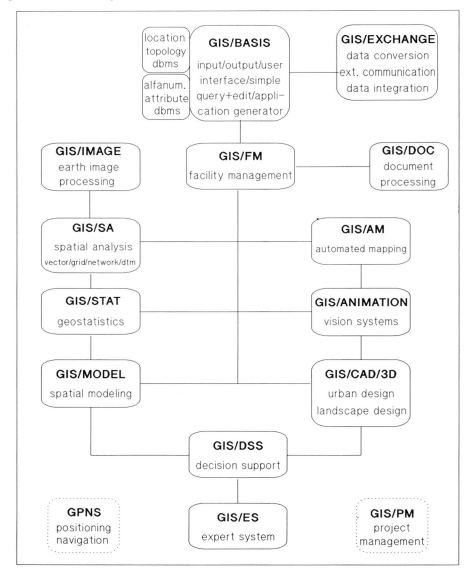

Figure 2.4: The family of geographical information systems

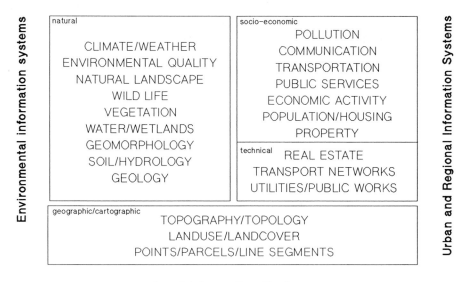

Land Information Systems

2.5 Problem areas of GIS use

Geographical information systems are not the first automated information systems that have become available for urban and regional planning. During the late 1960s and early 1970s, statistical and spatial modelling programmes were introduced, which made use of main-frame computers with batch processing procedures. Sauberer (1987) states that in this pioneer phase problems and misunderstandings appeared between the constructors of the hard- and software on the one hand and the planners and researchers on the other. Eventually this has led to a more or less complete abandonment of computer applications in urban and regional planning. The main reasons for the conflicts were as follows (Sauberer 1987):

(i)   the reluctance to adopt an objective-analytical approach in planning;
(ii)   the necessity to use large amounts of data, which could not be adequately handled by the first generations of computers;
(iii) the impossibility of these computers to process graphical map information;
(iv) the lack of training of planners and researchers in the field of automated information processing; and
(v)   the large investments and operational cost of hardware and software.

At the moment the prospects are brighter. A number of the points Sauberer made are no longer applicable. There are now much better and cheaper computers. They are capable of processing graphical data. System and application software has become more user friendly and there is now a general opinion and attitude which favours the full utilization of the new information technology.

Breheny (1987), however, observes that the development and use of the new generation of information systems emphasizes spatial management applications above the support of policy-making. Instead of supporting and enhancing the analysis and understanding of the environment, the emphasis with the application of computer systems has shifted towards increasing the efficiency and effectiveness of the organization. This distinction between the functionality for spatial planning and for spatial management is crucial when the usefulness of GIS in urban and regional planning is assessed. Partly because most efforts have been made to improve the management capabilities of GIS, some specific needs and wishes of planners and researchers are still not satisfied by the present-day systems:

(i)   Geographical information systems work in absolute and concrete space. Urban planning, however, operates at the intersection of social space and physical space. In social space, abstract, subjective and relative notions of space are of vital importance, but they are difficult to deal with in a GIS. Furthermore, information on individual spatial behaviour, a major theme in planning research, cannot easily be stored, managed and analysed with a GIS.

(ii)   Planning activities are often of a non-routine character and the policy-making process knows many irrational moments. The planning issues change according to prevailing political priorities, and procedures are also relatively easily altered. This has to do with the succession of 'planning cultures' mentioned earlier. All this requires a level of flexibility that most information systems cannot yet reach.

(iii) Planning involves much more than automated information processing. This means that planners only seldomly use information systems intensively and on a daily basis. The GIS systems that are available at the moment, have still to be operated by information specialists or 'technical' researchers. Middle management and more qualitatively oriented researchers already have problems when using the systems and for policy managers and incidental users the systems are too complicated to operate by themselves. Geographical information systems should also be better connected to or integrated with other information systems, such as document processing systems and project management systems. This will increase the possibilities of being able to use GIS for procedural planning purposes too.

(iv)   The problem of how to extract research, planning and policy information for decentralized processing from large central graphical and attribute GIS databases that are primarily used for registration, administration and management purposes, has still not been solved properly.

(v)   The costs of training, support, customizing and the building and maintenance of central databases are often still greatly underestimated in information plans and estimates of expenditures for information projects.

2.6   Possibilities for GIS applications

There are many possibilities for application of the GIS technology in urban and regional planning. This is especially the case for the substantive part of the planning activities. In order to structure the discussion, the planning process is broken down into elements (Figure 2.5). From a broad perspective on planning, analytical, policy-making and management tasks can be distinguished.

With respect to background studies, GIS can be employed for nearly all research that involves land based spatial analysis and modelling. Especially for area monitoring (both on a sectoral and integral basis), regional potential and feasibility analyses and site selection studies, the present-day GIS systems offer good functionality. Better possibilities for working with distance cost surfaces, measured along multi-mode transportation networks, would however greatly enhance many analyses. In the field of forecasting (simulation) there is still a lack of adequate tools.

At the moment, the presentation of integral cartographical views of strategies, plans and programmes is probably the strongest contribution that a GIS can make in planning studies. Both for policy makers and for the general public, GIS maps can be very effective means of information transfer. For studies in which plan alternatives are generated, much more flexible design, optimalization and evaluation tools would be needed in order give GIS a dominant position in the development process. Operations research methods and techniques are now more or less absent in standard GIS packages.

Figure 2.5: Elements of the urban and regional planning process

background studies	urban/regional monitoring	sector studies
(analysis)	problem/potential analysis	locality studies
	urban/regional forecasting	integral studies
planning studies	strategy/plan generation	
(policy-making)	plan presentation/evaluation	
plan implementation	project generation/approval	
(management)	subplan generation/approval	
	plan monitoring	

When a plan is adopted it has to be implemented. In this operational phase of planning, often new, more detailed plans will be made or development projects are started. If these plans and projects require the processing of georeferenced information, again GIS can be used in many cases. GIS can also be helpful for the documentation of spatial plans and in the approval process for development, building and installation permits. Applications can relatively easily be compared with the goals and the regulations of the plan. Also, information can be given to the public and to private firms about opportunities offered in the plan. Finally, the plan itself can be monitored in order to regularly assess the need for a revision.

2.7 Some outlooks for the near future

The GIS technology is still rapidly evolving. A number of the deficiencies of the present-day systems will certainly be cured in new generations of geographical information systems. With the trend towards the use of graphical UNIX workstations for GIS, the installation of clusters of workstations interconnected by advanced data communication networks, together with the introduction of servers on which the central databases and the network are managed, will allow for efficient distributed and dedicated information processing. Processing power, storage capacities and high resolution colour-graphical output facilities will all improve considerably. Total costs for automation will, however, hardly drop any further as falling hardware prices will be offset by the installation of more extended systems. Software, training and maintenance costs can be expected to remain relatively stable.

GIS systems will become more easy to operate and will be increased in functionality. A general introduction of graphical user interfaces and object oriented approaches to databases is very likely. An object oriented GIS environment will allow for the definition of non-spatial feature relationships above the locational relationships represented in the topology of the graphical database. In this way, geographic features can be represented in a more realistic model of reality. This will enhance the possibilities for sophisticated queries, analyses and modelling based on both values of spatial and attribute characteristics and on relationships of real world objects.

The conclusion from this short overview is that GIS can already play an important role in urban and regional planning, and that still some severe limitations exist with respect to the technical functionality of the systems. A number of these deficiencies will be overcome in the next decade and GIS is well on its way to acquiring the same status with respect to spatial information as word processors have for text and statistical packages have for number sets.

References

Braedt, J. (1988) Satellitenbilder als Informationsquelle für Landesplanung und Umweltschutz, Mitteilungenblatt, Deutscher Verein Vermessungswesen, Landesverein Bayern, 40, 117-131

Breheny, M.J. (1987) Information systems and policy formulation: changing roles and requirements, in Spatial Information and their Role for Urban and Regional Research and Planning, Bundeskanzleramt, Wien, pp. 25-41

Cammen, H. van der (1980) De sociale wetenschappen in het ruimtelijk beleid: twee opvattingen, in Opstellen over planologie en demografie, Planologisch en Demografisch Instituut, Amsterdam, pp. 61-77

Dangermond, J. (1988) GIS trends and comments, ARC News, 13-17

Kreukels, A.M.J. (1985) Planning als spiegel van de westerse samenleving: de frontline van de nieuwe grootstedelijke plannen in de Verenigde Staten in de jaren tachtig, Beleid en Maatschappij, 12, 311-324

Sauberber, M. (1987) Some requirements on spatial information systems at the national level, with emphasis to new planning strategies: the situation in Austria, in Spatial Information Systems and their Role for Urban and Regional Research and Planning, Bundeskanzleramt, Wien, pp. 13-16

Wissink, G.A. (1986) Handelen en ruimte: een beschouwing over de kern van de planologie, Stedebouw en Volkshuisvesting, 67, 192-194

Henk F.L. Ottens
Rijksuniversiteit Utrecht
Geografisch Instituut
Postbus 80115
3508 TC Utrecht
The Netherlands

3 GROWTH OF GEOGRAPHICAL INFORMATION SYSTEM APPLICATIONS IN DEVELOPING COUNTRIES

Michael F. Nappi

3.1 Introduction

The growth of Geographic Information Systems (GIS) in developing countries is linked to the relationship between driving and enabling forces. Driving forces are defined as those interrelated factors which, through their own development, create a set of circumstances requiring increased management and attention. Enabling forces, on the other hand, are those which supply the technology required to manage and observe the driving forces. GIS is becoming the dominant application area within the overall mapping market as more users require systems to analyse data. There still exists a number of critical factors which will determine the degree of further growth: cost, distributed systems, user interfaces, data integration and system support. Much of this growth in the use of GIS has become possible through the support of international agencies such as the World Bank and the U.S. Agency for International Development. Developing countries which utilize GIS include Egypt, Puerto Rico, Costa Rica, Venezuela, Colombia, Uruguay and Chile.

Geographic information systems are generating a great deal of interest worldwide as organizations become aware of the technology and its benefits. A GIS can be defined as "a computer based technology composed of hardware, software and data used to display and analyse geographic related information" (Dataquest 1988). These systems were developed as a method to integrate and analyse different data types into a single map in order to summarize geographic, socio-economic and various other attributes. Utilization of GIS has been growing rapidly worldwide, primarily due to the interest from a diverse set of users from agencies as identified in Figure 3.1.

Figure 3.1: Potential GIS users worldwide

Government	**Private Companies**	**Utilities**
Land management	Oil/gas companies	Electric
Geological survey	Forestry	Telephone
Defense mapping	Mining	Gas
Tax assessment	Minerals	Water
Transportation	Environmental	
Public works	Marketing	
Military	Distribution	
EPA		
Law Enforcement		
Emergency		

H. J. Scholten and J. C. H. Stillwell (eds.),
Geographical Information Systems for Urban and Regional Planning, 23–30.
© 1990 *Kluwer Academic Publishers. Printed in the Netherlands.*

The overwhelming amount of data these types of agencies have collected over time can often provide significant value, as long as there is a structured way to analyse this wealth of information. A GIS can supply this capability in those instances where the information is geo-referenced. Government agencies and utilities are benefiting from GIS worldwide as the technology is being made available to a larger audience each day.

The application of GIS in developing countries does vary from country to country. The underdeveloped countries are driven to establish land administration based systems by those entities supplying the funds, such as the World Bank and the U.S. Agency for International Development (U.S. AID). In other countries, the desire for productivity increases and cost reduction is driving the use of GIS. Elsewhere, GIS is being used as a decision support system based on land information which is a key to economic development.

3.2   GIS developments

In order to understand the growth of GIS in developing countries, it is important to understand how GIS has progressed as a technology. Computer based mapping, and more currently GIS, is a computation intensive activity which has required powerful systems capable of analysing diverse data sets and graphically displaying the results. The computers capable of handling these applications were mainframes in the 1970s and are now the powerful emerging technical workstations of the late 1980s (Table 3.1).

Table 3.1: GIS trends (modified from McLaughlin 1988)

	1976	1981	1986	1991
System Architecture	Mainframe	16-Bit Mini	32-Bit Mini	32-Bit Micro
Processing Environment	Central D/P Dept.	Dedicated Processor	Professional Workstation: Host-Based Network	Personal Workstation Distributed Network
Primary Emphasis	Map Production	Mapping and Attribute	Geographic Modeling	Decision Analysis
Software Cost	Non-commercial Software	$100 to $200K	$100K to 200K	< 100K
Hardware Cost Cost (CPU, Storage, input and output)	Approx. $450K for Mainframe Alone	Approx. $150-$300K PDP-Based System	Approx. $100-$250K Vax-Based System	50K and Less

While technology has grown in terms of power and capability, the role of computer mapping systems has gone from one of just generating cartographic maps to one of aiding in decision analysis based upon thematic mapping. In conjunction with these changes, the cost of mapping systems has fallen from millions of dollars to thousands of dollars which in turn has enabled a broader group of individuals to take advantage of the technology. Another area which has aided in rapid acceptance of this technology has been in the recent work accomplished in user interfaces and relational databases. These developments have allowed for user interfaces to be built to geographic information systems in order to address the interrelated data specific to the task at hand. Interfaces of this type can be customized for a particular application, such as telephony, and thereby allow for the user to query the database in terminology familiar to that line of work.

As with any technology, there are driving forces pushing the need for the system, and enabling forces, which make the system available in a useable and cost effective fashion. Driving forces can be defined as those interrelated factors which through their own development create a set of circumstances requiring increased management and attention. Driving forces include population growth, economic development, urban growth, agricultural development, natural resource management and land management.

Enabling forces, on the other hand, are those which supply the technology required to manage and observe the driving forces. Examples of enabling forces include powerful emerging workstations, commercial relational databases, lower cost GIS software, educated users, funding assistance and user interface developments. It is the combination of the two which drives the rapid growth of a particular technology. One without the other does not lead to the situation that would promote the creation of a specific solution, such as GIS.

The sheer growth of the world population (Figure 3.2) by itself creates a set of circumstances requiring management (e.g. urbanization and world food supplies). World food supplies in turn require the management of land and the observation of weather patterns. This brief example shows one of many interrelationships important to developed and developing countries which can be managed through the use of a GIS. Of course, these relationships have been important for a number of years, but it is the availability of enabling forces, such as technical workstations, which makes GIS technology useable by an increased number of countries to manage the ramifications of these interrelationships.

Although computer based mapping began in the 1960s, the technology to support true geographic information systems is only just emerging. The technology advancements important to GIS use in developing countries include technical workstations, spatial and attribute relational databases, optical storage, commercially available GIS and increases in available relevant data. It has only been in the last few years when off-the-shelf GIS solutions, rich in features and functions, started to become accessible to users throughout the world. In support of these solutions have come increases in readily obtainable data from sources such as satellites and government developed databases.

One advancement which has played a significant role in the expansion of GIS is the recent increase in power at the desktop. Technical workstations, and even personal computers, have benefited from advances in computer chip research. Today's personal computers, utilizing true 32-bit processing chips, have performance ratings in excess of the mainframes which were used only a decade ago. The technical workstations based on the Reduced Instruction Set Computing (RISC) architecture available today even outperform these personal computers by at least fourfold. However, it is not just the power, but more so the drop in cost, which has enabled the proliferation of desktop mapping. Every year more power is available to the average user at a constantly reducing cost which encourages the use of desktop GIS solutions.

The continuous growth in world population and the availability of cost effective workstations, two seemingly unrelated events, do interact when we address the management of today's world resources. Figure 3.2 shows the relationship between population growth and computer power, represented in Million Instructions Per Second (MIPS). The graph depicts how population growth and the resulting increase in data associated with increasing populations drives the need to manage more information. When a supply of the technology required becomes available, represented by the increase in MIPS, a situation is created that enables increased use of systems required to administer growth dependent factors. This type of relationship is important in the growth of many activities associated with the day to day regulation of urban growth and resource management. Computer resources once available only to the larger and wealthier countries in the world are now becoming obtainable to even the poorest of nations.

Figure 3.2: The growth of driving and enabling forces

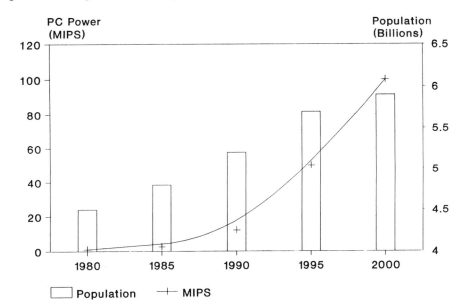

There still remain a number of critical factors which will determine the degree of further growth in the use of GIS throughout the world:

(i)   cost: users require low cost, powerful systems for the majority of GIS applications;
(ii)   distributed systems: users have the requirement for distributed data and resources across organizations;
(iii) user interfaces: the largest share of potential users are not computer experts and therefore require easy to understand application tailored interfaces;
(iv)   data integration: solution systems require the capability to merge information composed of both raster and vector data; and
(v)   system support: many users require organizations to provide training, hardware and software support and customization.

In addition, many of the systems being used to manage resources today have user interfaces written in English only and are not easily nationalized for a specific country's language. Furthermore, the support organizations necessary to help build skills in GIS are based in Europe and the United States for the most part, prohibiting use in developing nations. This impacts growth of system implementation because there are few individuals knowledgeable in the set up of these sophisticated networks and workstations with respect to GIS. Another critical factor is the lack of quality data to be analysed by a GIS. Even with satellite data available on a wider scale, it is the need for other data types, such as aerial photography and the establishment of geodetic control points, that truly slows the implementation of GIS in developing countries. The combination of these factors will impede the growth of modern GIS technology until truly international organizations begin to supply what is required on a worldwide basis.

3.3   Worldwide GIS activity

In the past few years, the supply of GIS technology has grown for developing countries. With assistance from organizations such as the World Bank, U.S. AID and others, less developed countries have been able to implement GIS solutions to handle problems specific to their needs. Programs have been put in place to manage many diverse applications, such as: global deforestation, urbanization, soil erosion, mineral and petroleum exploration, weather trends and many more. U.S. AID, for example, has been helping many countries with monetary assistance for development projects and many of these have implemented GIS.

In fact, the majority of developing countries are either investigating or implementing GIS solutions today. Driving forces in these countries are coming from the requirement to manage urban and economic growth without sacrificing their own natural resources in lieu of growth. Although it is not in the scope of this chapter to list all GIS activity in developing countries, some examples are as follows:

Puerto Rico

The Telecommunications Authority and the Property Taxes Office of the Treasury Department have been evaluating GIS solutions to their particular problems. The Telecommunications Authority is wanting to develop an application to manage their outside plant activities. The Treasury Department needs to develop a cadastral database that will be used by tax assessors.

Colombia

In the City of Medellin, the Public Utilities and the Property Tax and Metropolitan Planning Divisions are currently investigating the development of a multi-purpose GIS. The Public Utilities will use it to regulate water, waste disposal, energy and telephone services to the city. The Property Tax and Metropolitan Planning agencies will build a database to maintain cartographic data plus all land use related information.

Egypt

Possibly the largest mapping project in the world has been initiated by U.S. AID to monitor irrigation flow in the Nile Delta, involving the mapping of over ten million parcels of land and the inclusion of associated data into the land information system.

Costa Rica

The National Cadastral Office intends to develop a cadastral database · the densely populated central plateau of Costa Rica. The main applications ·ʳ GIS will be for tax assessment and land ownership recording.

Venezuela

The National Water Authority is evaluating the use of GIS to develop a database of their pipe network. This database will aid the Water Authority in the maintenance of the existing network and also in the planning of new pipelines.

Uruguay

The Department of the National Cadaster will use a GIS to develop a base map with a total of 1.2 million parcels, all the parcel attributes and all city block attributes.

Chile

A project is underway in the City of Santiago to develop a GIS application for the electrical distribution network in the city. The application will be designed to recognize network connectivity and flow.

The significance of these examples is that most countries are starting with pilot systems based on technical workstations, which would not have been possible until relatively recently. Systems of this type are either stand alone in nature, or are networked to file servers and mainframes for data access. The point is that the majority of the analysis is now done locally instead of on the host system.

This short synopsis of activity in developing countries shows the different ways governments and local agencies are beginning to manage the large quantities of information and resources through the use of GIS systems. Activities of this kind are being replicated throughout the world as increasing numbers of countries realize and take advantage of the benefits of geographic information systems.

3.4   Future GIS growth

Even with the present flurry of GIS activity being generated there is every sign that this is just the beginning of what is to come. Recent forecast projections show the overall use of computer mapping systems growing significantly over the next few years with large growth expected within North America, Europe and the Far East (Figure 3.3). Dataquest Incorporated, which monitors the mapping market, suggests the international markets will be driven by the need to manage growing infrastructures whereas North America will begin to take advantage of the desktop mapping systems. This scenario of first managing a country's infrastructure and then moving to desktop mapping, will more than likely be the path most countries follow as the supporting network for GIS is established. The network, composed of the aforementioned enabling forces, allows for the individual to take advantage of GIS.

Figure 3.3: Dataquest's worldwide mapping forecasts by major world region

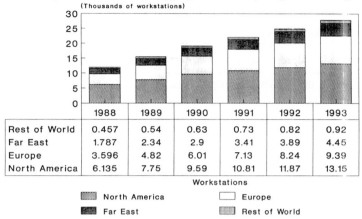

	1988	1989	1990	1991	1992	1993
Rest of World	0.457	0.54	0.63	0.73	0.82	0.92
Far East	1.787	2.34	2.9	3.41	3.89	4.45
Europe	3.596	4.82	6.01	7.13	8.24	9.39
North America	6.135	7.75	9.59	10.81	11.87	13.15

Workstations

North America   Europe

Far East   Rest of World

The major share of growth in the overall mapping market will be in the GIS sector, with GIS becoming the dominant mapping technology in the 1990s (Figure 3.4). This implies that systems which only create maps will become replaced by geographic information systems that are richer in features and functions required to analyse the impact of growth and change. These situations are taking place in many areas within application software development today. The drive is to supply the total solution and not just a distinct function that serves limited purposes. Developing countries will require more features, not less, as worldwide growth demands the making of more informed decisions.

Figure 3.4: Dataquest's worldwide forecasts for total mapping and GIS

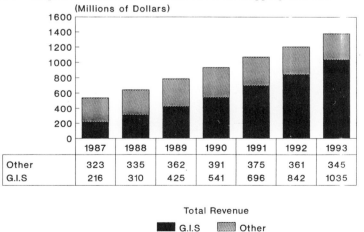

	1987	1988	1989	1990	1991	1992	1993
Other	323	335	362	391	375	361	345
G.I.S	216	310	425	541	696	842	1035

Total Revenue

G.I.S   Other

3.5   Conclusion

Geographic information systems are becoming widespread in their application by developing countries worldwide. Solutions once only capable of running on mainframe computers are now being implemented on stand alone technical workstations due primarily to the advances in computer performance. The stride made in computer power at the desktop is just one of many important enabling forces for this technology. The need for GIS is driven by factors such as population growth and urbanization, which in turn create various types of geo-referenced data. Information of this kind lends itself well to the analytical capabilities of geographic information systems. Additional uses will arise throughout the world for these systems as users are educated and data becomes available in useable formats. GIS utilization will continue to expand worldwide as global environmental problems and population growth drive the need to manage the earth's resources.

Acknowledgements

The author wishes to acknowledge the following Unisys colleagues: Johan Norvick, Suresh Sood and Peter de Gouw for the information supplied on worldwide GIS activities and Paul Hall for his help with the graphics. The author would also like to acknowledge Karen Vogel and Pat Ellis for their editorial review.

References

Australasian Advisory Committee on Land Information (1988) Status Report on Land Information Systems in Australasia 1987-1988: Canberra, Goanna Print
McLaughlin, J. (1988) Geographical Information System Concepts, Proceedings of the GIS Seminar in Toronto: Queen's Printer of Toronto, pp. 10-25

Michael F. Nappi
UNISYS Corporation
PO Box 500
Township Line and Union Meeting Roads
Blue Bell PA 19424
USA

PART II
DATA MANAGEMENT

4   INTELLIGENT INFORMATION SYSTEMS FOR ACCESSING PLANNING DATABASES: THE SAN FRANCISCO EXPERIENCE

Poulicos Prastacos and Pertti Karjalainen

4.1   Introduction

Planning techniques and procedures at the regional and urban level have been hindered in the past by the lack of available information which could be used to guide decision making. Whereas for many developing countries this problem arises because regional data sets are scarce and incomplete, in several developed nations, there is a wealth of regional and urban data. Planners, however, have not been successful in using these for everyday decisions. The main reason for this has been the lack of information systems that can assist users in accessing and analysing information in an efficient manner.

Urban and regional data must be organized in computer databases if they are to be used efficiently. Vast amounts of information make it impossible to analyse data available in printed form only. Unfortunately, computers have not been used extensively by the planning profession in the past. During the 1970s and early 1980s for most of the planning organizations in the United States, computing has taken the form of a 'central' mainframe based facility used mainly for financial and accounting purposes. Occasionally, these computers were used also for transportation modelling, but the difficult operating system language made them accessible only to a small elite of users who possessed extensive computer expertise. With the recent proliferation of microcomputers, planners finally have the opportunity to analyse the available information using tools provided by modern information technology. Commercially available relational DataBase Management Systems (DBMS) can be used to organize databases on microcomputers. Most important, special software can be written as a front interface to the database so that the inexperienced user can be sheltered from the intricacies of the database itself (organization, contents), as well as the DBMS used for retrieving the stored data. Along these lines Scholten and Meijer (1988) have developed RIA a user-friendly regional database system, whereas others (Lorie and Meier 1984, Rhind 1987, Openshaw and Mounsey 1987, Stephenson 1989) have discussed the application of advanced information techniques in regional and GIS databases.

This chapter outlines two intelligent information systems developed in the San Francisco region in the USA. The CeDAR (Census Data Analysis and Retrieval system) was originally developed to access socioeconomic data from the 1980 Census of Population but it can be used to access many other different sources of regional data. The Marin Countywide Network System (MCNS) is currently being developed to permit cities in Marin county to access and update parcel information in the County Assessor's Master Property File. In both of these systems the emphasis has been placed in providing a system that computer illiterate users can easily learn and utilize in the course of their everyday work.

4.2   Database management systems

A DBMS is software that permits the storage, retrieval and management of information in an

H. J. Scholten and J. C. H. Stillwell (eds.),
Geographical Information Systems for Urban and Regional Planning, 33–42.
© 1990 *Kluwer Academic Publishers. Printed in the Netherlands.*

orderly manner. DBMS systems have been extensively used in the private sector for a variety of operations (management information system, decision support).

Relational DBMS

The widespread use of DBMS has been encouraged by the commercial availability of relational DBMS packages for mainframe computers. The concept of organizing information in a relational database was first introduced in the early 70s (Codd 1970), but it was not until the late 70s that the first relational DBMS became available for mainframes (Martin 1986). Relational databases are more flexible than their hierarchical or network counterparts since they represent data as two dimensional tables, referred to as 'relations' (Date 1981). Data are free of path or sequence dependencies and users can easily manipulate the database to define new 'relations' as they choose. The concept of a relational database can be readily applied to the data we use in regional planning. Any collection of data sets can be classified along three dimensions:

(i)   spatial level (nation, provinces, counties, cities, census tracts, etc.);
(ii)   variables or attribute data (various population/economic/environmental characteristics);
(iii) time (year, month, day).

A relational database from this three dimensional data set is formed by considering the spatial level and the variables dimensions as the two dominant dimensions of the database, while the time dimension is mapped to the spatial level dimension through appropriate algorithms.

As a result of the success of relational DBMS in the mainframe world the vast majority of the DBMS currently available for microcomputers are of the relational type. These systems offer most of the capabilities of their mainframe cousins in the user-friendly environment of microcomputers. They can be used to organize and manage regional databases and, most importantly, contain their own programming language in which special applications can be written. This facilitates considerably the development of user friendly information systems since it permits the design of higher level query languages which offer powerful operations without demanding explicit knowledge of the programming concepts of the operation of the DBMS.

There are several DBMS packages for microcomputers. The information systems discussed in this chapter have been developed using the dBASE III+/IV (tm) command language. dBASE is probably the most widely used DBMS and contains a powerful programming language. The success of dBASE has fueled the development of various dBASE-compatible, add on software packages which enhance and extend the dBASE language. Two of the most prominent of these are the FoxBase (tm) DBMS and the CLIPPER (tm) compiler. The latter is a compiler; it can read programs written in dBASE command language and compile them to produce executable (EXE) files. Since compiled programs usually run as much as 10 times faster than non-compiled programs and users of the program cannot alter the source code, most of the commercial applications of dBASE, CeDAR and MCNS included are distributed after being compiled with CLIPPER.

Intelligent information systems

Organizing the available data in a computer database is not necessarily going to lead to widespread adoption of the database for planning purposes unless the user is offered an easy-to-use mechanism to access the information. Users of the databank must know several things before they are able to efficiently use the database. They must know:

(i)   the organization of the database (structure, field names, relationship etc.);
(ii) the appropriate commands of the DBMS for accessing and retrieving the data; and
(iii) the contents of the database, that is the data items stored, and their meaning.

The first two items require computer expertise which most planners do not possess. To become proficient in the advanced DBMS used today requires extensive training and often knowledge of how computers operate. DBMS user manuals are voluminous and difficult to use. Additionally, users of the database must be provided with enough information on what data items are contained in the database. In CeDAR there are a total of 1300 different data items; it is practically impossible to memorize the various data variables.

One way to overcome these problems is by developing an interface, a front-end, to the database whose function is to assist users to access and analyse information in an efficient manner. This user interface provides a 'human window' to the system and handles the interaction between user and data. It transforms the information system from a mere database/bank to an intelligent information system. A well designed user interface can relieve the user from the burden of learning the programming/query language to access the information or the format/organization of the database. By employing menus or some query language familiar to planners the various options are presented to the user in a well understood manner. User's requests, expressed in natural language or solicited through structured questions are translated by the user interface to DBMS instructions which are then executed. The front-end to the database must also provide the role of the 'navigator' through the database. It should present the user with the set of available data variables and their meaning. An on-line dictionary or table of contents, appropriately indexed and/or organized can assist users in browsing through the database.

4.3   CeDAR

Background and description

The Association of Bay Area Governments (ABAG) is the regional planning agency of the San Francisco region in California. Two important functions of the agency are to analyse for various planning purposes the data published by the Bureau of the Census and to disseminate information on local socioeconomic characteristics to other government agencies and decision makers in the private sector. An elaborate information system was developed in the early 80s. This system was operating on VAX minicomputers and included several databases the most important being the results of the 1980 Census. In early 1985 it became clear that this system, although quite elaborate, could not support the functions for which it was originally designed. Planners without the expertise to operate minicomputers had to depend on the data processing department for any type of analytical function. This created a bottleneck and resulted in underutilization of the information wealth that can be found in the Census data. Additionally, ABAG could provide outside users only with printed reports which, although adequate for most users, were of little use to decision makers that desired to analyse a large subset of the Census data.

It was for these reasons that CeDAR was developed. CeDAR is a microcomputer based information system that can be used to store, retrieve and analyse regional information (Prastacos and Karjalainen 1989). The system was written in dBASE III+/ CLIPPER (tm) command language and presently contains two databases, data from the 1980 Census of Population and ABAG's projections of future economic and demographic conditions in the region. Through CeDAR planners are able to analyse the census information on their desks without the help of the data processing people. CeDAR functions not only as an avenue to distribute information on local socioeconomic characteristics to planners in the private sector but also, provides them with a tool to analyse these information sets and therefore make more intelligent decisions.

CeDAR was designed for users with minimal previous experience in computers. Its structure and its capabilities reflect the daily work performed by planners. Today, it is fully operational and is used by over 25 different organizations. The data sets contained are quite large; the Census database includes 1200 data items (attributes) for each one of the 1500 census tracts and cities of the region while the Projections database contains another 60 attributes for the same areas. There is a total of about 2 million data items that users can select. Each one of the variables has been assigned a short mnemonic name which users specify to retrieve and display information. For example, population is referenced as 'pop', whereas, for the number of households the standard abbreviation 'hh' is used. Areas are referred to with their English and/or 'planning' names rather than a code. Thus, to retrieve data for the city of Berkeley, users type 'Berkeley' and not the Bureau of the Census code number corresponding to Berkeley.

Data dictionary

To assist users in finding the variables contained in the system and their mnemonic names, a complete data dictionary is included on line. The data dictionary has a hierarchical form; the 1200 different variables are grouped to 140 tables which are further grouped into 24 major categories. Thus, the major category of 'Income' contains several tables ('Household Income Distribution', 'Family Income Distribution', etc.) while the 'Household Income Distribution' table contains 19 variables, the number of households at various income levels. Using this structure the user of the system can easily focus on the data items of interest.

Analytical capabilities

Since CeDAR was designed to assist planners in their everyday work it has certain analytical capabilities (Table 4.1). Firstly, there is the possibility of estimating user defined variables. Quite often planners are interested in data items which are algebraic transformations of variables contained in the Census database (percentages, ratios). This can be accomplished in CeDAR by typing the names of the appropriate census variables separated by arithmetic operators (+, -, *, /). The software would then perform all the necessary data retrieval and mathematical operations and report the results for the areas selected. Secondly, areas can be aggregated. Analysis and planning does not always occur at the spatial levels at which the Census reports information. For example school or health districts and traffic zones consist of more than one census tract. Users of CeDAR can define the tracts to be aggregated and the system will estimate the variables for these new areas.

Thirdly, the package allows the user to perform conditional searches. CeDAR has the capability to search the complete database and retrieve information only for areas that satisfy a condition the user specifies. With this feature of CeDAR questions of the type 'which tracts have a median household income greater than $30,000? ' or 'which areas in the city of Berkeley have a large percentage of households with income below poverty level ?' can be answered with a few keystrokes. Conditions can be specified using any relational (<, >, =, <, >) and/or logical operators ('and', 'or').

Interaction with spreadsheets, and mapping programs

CeDAR provides users with the choice of displaying the data on the screen, printing them on the printer or writing them in a disk file for later processing. There are special facilities that permit users to create disk files which are then directly readable by spreadsheet (LOTUS) programs. ASCII files can be also created which can be then imported to other DBMS. Often the results of an analysis can be more easily visualized when displayed on a map that shows the boundaries of the different areas rather than a table that has only their names. CeDAR has facilities to create files which are readable by two mapping packages, ATLAS (tm) and MapInfo (tm). ATLAS is a mapping software which can produce thematic maps. To use CeDAR and ATLAS together, users create with CeDAR a specially organized disk file with all the necessary information and then ATLAS reads this file to produce the map.

Table 4.1: Analytical capabilities of CeDAR

```
Existing Variables

    POP = Population
    BLACK = Black Population
    HH = Households
    HI_MED = Median Household Income
    WHITE = White Pop.

New Variable Creation

    POP / HH                Household Size
    100 * (WHITE + BLACK)   Percent of the Population which is White or Black

Area Aggregation

    TR 4001; 4005;          Aggregate tracts 4001, 4005 and
    4008 - 4018             4008 thru 4018
    PL BERKELEY;            Aggregation of cities of Berkeley and
    OAKLAND                 Oakland

Conditional Search

    Find areas that satisfy the following conditions:

    POP > 2000 AND          Find areas with population more than 2000 and
    HI_MED > 30000          median household income more than 30000

    (POP / HH < 1.6) AND    Find areas with household size less than 1.6 and
    HI_MED > 30000          median household income more than 30000
```

The interaction of CeDAR with MapInfo which is currently being developed is more interesting. MapInfo is a powerful mapping software that can display streets and boundaries and can read directly dBASE III+/IV files. Associated with MapInfo is MapCode, a programming language to write MapInfo applications that can be compiled and linked to other programs. With the MapCode language, CeDAR and MapInfo can be combined into one software. Users can display data in thematic maps without having to exit from CeDAR. This connection of CeDAR and MapInfo transforms CeDAR to a pseudo-GIS and greatly enhances the utility of CeDAR.

4.4   The Marin countywide network system

Description of the system

Efficient countywide planning and permit-tracking is often made difficult if not impossible by the lack of centralized, up-to-date data. Data exchange between county government on one side and city/town governments on the other is, in many cases, very slow and incomplete, resulting in improper property assessments and long delays in permit approvals. Until very recently, this was the case in Marin County, which is located immediately north of San Francisco. Local planning agencies were able to browse through countywide parcel data only on microfiches, which were updated once or twice a year by the Assessor's Office. On the other hand, the County Assessor's Office and planning agency became aware of new developments at the city level long after they were already completed. As a result, supplemental tax assessments were delayed for long periods of time, costing the cash-strapped county a lot of money, and the central planning agency did not have a clear, up-to-date picture of the most recent developments in the county, which made efficient planning quite difficult.

It was mainly for these reasons that the Marin County Planning Department decided to create a PC-based countywide network system between county and city planning departments. The main goal of this system was to provide an inexpensive, central database retrieval and update system for all planning and permit data in the county. The core of this database consists of the County Assessor's 'Master Property File', which includes information on owners, assessed value of land and improvements, new building permits etc. for all parcels within the county. This data is periodically downloaded from the assessor's minicomputer to a fast 386-PC connected to several telephone lines. Local planning departments dial into this 'server' via modems, and may either download datasets to their own PCs or search, analyse and edit data directly on the server. Once or twice a month the changes made by local jurisdictions to the master property file are reported and merged to the main database on the county government's minicomputer.

The benefits of this on-line database can be tremendous for both central and local planning agencies. The County Planning Department and Tax Assessor's Office will become aware of any changes on the status of any given parcel as soon as they are recorded at the local level. This way the Assessor's Office can order property reassessments and send supplemental tax bills as soon as (or even before) property improvements are completed. On the other hand, local planning departments can easily check the full status of any parcel within their jurisdiction, as well as, be able to see new and pending developments in neighbouring cities. This is becoming increasingly important for resource allocation and permit approvals. Although the data in the database is stored in four basic geographical levels - planning area, census tract, traffic zone and parcel, users can create and store their own areas and/or variables by combining existing data. For instance, a school district might consist of 800 parcels, and of particular interest to planners in that district might be the ratio of single family homes to multi-family dwellings. Planners can easily see this data by combining the appropriate parcels and dividing existing variables through easy and consistent menu choices. These new variable and area definitions can then be stored into a user-library for subsequent retrieval and the results displayed in any of the available formats.

It was recognized from the outset that such a system could be useful and efficient only if its user-interface was easy and intuitive and if data retrieval was relatively fast, in spite of the large size of the main database (over 40 MegaBytes). The database front end was designed to use stacked menus and "point-and-shoot" light-bars with on-line help available at all times. Data retrieval is speeded up by using various indexing-schemes and filters, and the user is given many output devices and options to choose from, including interfaces to LOTUS 1-2-3, dBASE and various other third party data analysis tools.

The MCN System is currently being used by almost all of the cities and towns in the county, and because of its ease of use and the value of its up-to-date data, it is being utilized extensively by local and central planning departments, as well as the Tax Assessor's Office for development-, application- and permit-tracking purposes.

Components of the system

MCNS consists of five basic modules:

(i)   Database selection;
(ii) Data retrieval and analysis;
(iii) Mailing label generator;
(iv) Electronic mail and bulletin board system;
(v)   System and data security.

Module 2 is broken down to six separate sub-modules:

- Area selection;
- Data selection;
- General area/data filter setup;
- New data/area creation;
- New data/area library management; and
- Report setup.

Database selection

MCNS users can choose to use data from many different databases. The number of available databases and data items (fields) and areas (records) within them is limited by system security, which restricts users' access to certain non-sensitive data and/or areas within the available databases. Database selection is done from a 'point-and-shoot' lightbar menu, and currently users can select only one database at a time (no relational/cross reference access). As soon as a database is selected, its supporting files (index, memo, etc. files) are automatically attached to the process and the database is initialized and filtered according to the user's security profile.

Data retrieval and analysis

Once a database is selected, users can proceed to select data and areas. These selections can be made either from menus by tagging desired items or by 'manual' definitions from the system prompt. Manual selections are used for instance when selecting a range of areas ('TRACTS 1011 TO 1056'). Selected data and areas can be further limited or filtered by queries of the type: 'TRACTS 1011 TO 1025 IN TOWN OF FAIRFAX, IF VALUE OF IMPROVEMENTS IS BETWEEN $100,000 AND $200,000'. In addition to selecting areas and variables that are explicitly provided in the database, users can create new areas and/or data by combining existing variables and records. For example, a planner might need to know the ratio of land value to value of improvements in a certain school district. Since this ratio is not part of the standard database setup, the planner would create the ratio by defining a new variable from existing variables as: 'LAND VALUE / VALUE OF IMPROVEMENTS'. He would then define the school district as an aggregation of parcels which belong to the district. These new variables and areas can then be given their own, user-defined names and report titles, and can be stored to area and data libraries for subsequent retrieval. After data and areas have been selected and/or created, a user can proceed to set up a report. Users can specify one of the following output devices: screen, printer or disk file; printer options include drivers for Epson/IBM-compatible dot matrix, as well as HP-LaserJet compatible printers. These drivers have options for typeface, margins, page length, line length, orientation etc. These selections are saved to a user profile file and automatically used as defaults in future runs. Output file options include 'print file' (with printer commands and proper formatting), comma-delimited ASCII (with separate structure/format specification file), LOTUS 1-2-3 and dBASE III/IV files.

Mailing labels

Local governments routinely mail a large number of notices and reminders to parcel owners. In most cases this is a very time consuming, manual process. The Mailing Label feature of MCNS was designed to expedite and automate this process, and it can be used to selectively send mail to any parcel in the database. For example, the database could first be filtered for only certain blocks within a city and the system would automatically print mailing labels for those parcels only. The selection of areas and/or global database filters for mailing labels is identical to area selections as outlined above. Once users has selected area criteria for labels, they can proceed to select one of many available label formats for Epson/IBM or HP LaserJet-compatible printers. The labels can be sorted on any combination of postal zip code, parcel owner's name and address and can be printed to screen, disk-file or printer.

Electronic mail and bulletin board system

Each user has his/her own 'mailbox address' within the system, into which other users can send messages and files. At the start of a session users are reminded of 'unopened' mail as long as there is new mail in their mailbox. Bulletin board systems work in almost exactly the same way as the electronic mail, except that all mail sent to the bulletin board is public and read-only, i.e. everybody can read it but only the sender of a message or the system administrator can edit or delete it.

System and data security

MCNS has a rigorous security system to protect the data from unauthorized access or modifications. The security system acts on both the operating system and application level. Each new user is given his/her own group assignment and personal, secret password, which restricts access to files, variables, areas and system functions. All changes to the databases are posted to a holding area, which is reviewed and approved by the system administrator prior to final merge into the target database. Changes to data are also stamped with date, time and the name of the user making the change. No user is given access to operating-level functions; users can either select functions from MCNS menus or log out of the system (hang up the phone).

4.5 Conclusions

CeDAR and MCNS are flexible and intelligent front-ends for large and complicated databases. Their main function is to assist planners and other data users in retrieving, analysing and updating large amounts of data in as easy and 'user-friendly' way as possible. They have been designed to be open-ended and modular, with clearly defined and easy 'hooks' for additional modules or completely separate third-party applications software. They are somewhat limited by the fact that they were written for PC-computers, but with the rapidly increasing computing speed and power of PC-systems, the performance of these systems is close to a state-of-the-art minicomputer of just a few years ago.

CeDAR and MCNS are true PC-applications in a sense that they utilize almost all of the modern user-interface features that microcomputer users have become to expect since the introduction of LOTUS 1-2-3 and its revolutionary command system. Also, since these applications read and write ASCII files, data exchange between them and various other microcomputer applications as well as mini- and mainframe computers is relatively simple. We perceive these data analysis systems to be forebearers of the next stage of microcomputer revolution, when the power and complexity and vast potential of microcomputers in data retrieval and analysis would require extensive training and expertise of their users in order to realize their full potential. CeDAR and MCNS are examples of distributed 'desk top' data processing systems. By taking the 'pain' out of using complex DBMS and hiding at least most of the machine interface behind an intuitive 'human' interface they can be easily used by anybody familiar with the basics of planning concepts.

References

Date, C. (1981) An Introduction to Database Systems, Addison-Wesley, Massachussets
Codd, E. (1970) A relational model of data for large shared data banks, Communications of the ACM 13, 377-387
Lorie, R. and Meier, A. (1984) Using a relational DBMS for geographical databases, GeoProcessing, 2, 243-257
Martin, D. (1986) Advanced Database Techniques, The MIT Press, Cambridge, Massachussets
Openshaw, S. and Mounsey, H. (1987) Geographic information systems and the BBC's Domesday interactive videodisk, International Journal of Geographical Information Systems, 1, 173-179
Prastacos P. and Karjalainen, P. (1989) An information system for analyzing census data on microcomputers, in Ekistics
Scholten, H.J. and Meijer, E.(1988) From GIS to RIA: a user-friendly microcomputer oriented regional information system for bridging the gap between researcher and policy maker, Paper presented in the 1st URSA-NET Conference, Patras, Greece
Rhind, D. (1987) Recent developments in geographical information systems in the U.K., International Journal of Geographical Systems, 1, 229-242
Stephenson G. (1989) Knowledge browsing: front ends to statistical databases. In Rafanelli, M., Klensin, J. and Svensson, P., Proceedings of the 4th International Working Conference on Statistical and Scientific Database Management, Springer Verlag, Berlin

Poulicos Prastacos
Association of Bay Area Government
PO Box 2050
Oakland
California 94604
USA

Pertti Karjalainen
Northern Lights Software
Fairfax
USA

5 GEOGRAPHICAL INFORMATION SYSTEMS DATABASE DESIGN: EXPERIENCES OF THE DUTCH NATIONAL PHYSICAL PLANNING AGENCY

Wouter M. de Jong

5.1 Introduction

At the Dutch National Physical Planning Agency (RPD), GIS technology has been used since the early eighties and has been incorporated more and more in the daily working practises. What began as the fancy tool of a number of specialists, is nowadays used throughout the RPD. One of the policy premises, as formulated in the Information Plan for 1989-1993, is that access to the 'Geographic Information System' is to be made possible for every researcher and planner as requested. In this chapter, the focus will be on database design and on the development of GIS at the Physical Planning Agency, in the light of present trends.

To give some insight into the use of GIS, and the different demands placed upon these systems, an overview will be given of the activities of the Agency, the information needs these activities generate and the existing geographical information systems. The third part of this chapter will be dedicated to resources and the efforts that have been invested in realizing the present availability of geoinformation. Section 5.5 highlights present work on developing a central geographical database, the considerations beforehand, the underlying concepts and some details of the project itself.

5.2 Activities at the RPD

The purpose of the RPD is to support the minister responsible for planning by formulating, implementing and controlling physical planning policy. To perform these tasks efficiently, the RPD is divided in several sections that concentrate on research activities, planning design activities, policy formulation and control of planning legislation (De Jong 1989b).

GIS and planning research

One of the major activities of the RPD is to compile information on all aspects of society on a locational basis. The purpose is on the one hand to generate a spatially differentiated picture of the state of society, and on the other hand to get a grip on trends in the development of spatial phenomena. The use of GIS technology in planning research therefore has several aspects. The flow of data and the processes that play a role are pictured in Figure 5.1 (see also Van der Schans 1988). Of course this is an idealized picture. Though reality is seldom to be captured in such a neat scheme, it gives an idea of the main activities.

Firstly, data have to be collected before being put into the GIS. All kinds of measurement technique are applied to gather data on socioeconomic, cultural and physical reality. Data collection techniques range from satellite remote sensing to interviewing people. Only rarely is the primary collection phase in the hands of the RPD. Far more frequently, data are supplied by other organizations. Sometimes special projects are undertaken in which, under direction of the RPD, private companies perform this task. The second phase is data processing in order to ensure the data fits the data definitions of the GIS. This phase may involve generalization,

43

H. J. Scholten and J. C. H. Stillwell (eds.),
Geographical Information Systems for Urban and Regional Planning, 43–56.

Figure 5.1: Schematic representation of activies at the RPD with respect to GIS

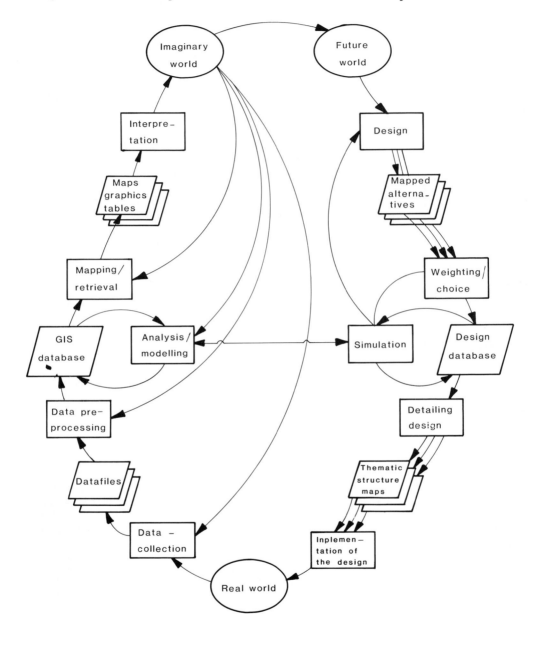

interpolation, resampling, transforming, rescaling, aggregation, and (re)classification of data. Processed data are put into the GIS database. In order to generate sensible images of the real world, various kinds of geographical and statistical analyses are necessary. To study a certain phenomenon, it may be that combinations of data layers are made (overlays), the extension of phenomena is estimated (e.g. by buffer or spread), time series are analysed to discover trends, and prognoses are made to predict future realities under certain conditions. In a GIS environment, results usually are presented on various types of maps. Finally, the resulting graphical images, tables or written reports are interpreted by researchers confronted with mental images or existing (theoretical) views of the real world. This confrontation may lead to a change in the processing algorithms described previously to correct output to be more in line with (theoretical) expectations. It might also induce a change in the view of the real world.

GIS and planning design

While planning research aims essentially to describe and comprehend real world issues by building encompassing databases and using algorithms, planning design actually works the other way around. Instead of adjusting the theoretical, mental images to the results of scientific analysis, designs of the desired future situations are placed upon present reality. The activities in planning design are therefore fundamentally different from those in planning research. Nevertheless, both may use GIS technology and both use abstracted images of reality. On the basis of knowledge of society's characteristics, conceived bottlenecks, development potential and trend analysis, design in physical planning is the translation of development goals into alternatives for the spatial arrangement of societal phenomena. In the Netherlands, these alternatives set limitations and potentials to constrain or sustain development trends. These alternatives develop in an interactive process of man and machine, whereby mental images are confronted and placed upon real world abstractions. Designs take the form of small-scale structure maps broadly indicating the desired real world situation. A further step might be the simulation of the effects of design alternatives on extrapolated data sets, and the estimation of their mutual influence. Given the results, a more profound weighting of design alternatives might be made. Such analyses are still in their infancy at the RPD. Once one of the alternative designs has been chosen, it has to be implemented. Thematic maps are produced that depict the direction of development for various indicators. At the level of national planning, these maps never give exact directions to local authorities on how to realize their plans, but they provide an indication of future investment preference and budget priorities.

GIS and planning policy

Part of the activities that have been described overlap with planning policy. While the generation of planning alternatives is pure planning design, the weighting of alternatives is partly the field of planning policy. In formulating new policies, it is not necessary to review the sifting of alternatives as done in the planning design section. However it is still necessary to offer variability to policy-makers. Political circumstances vary, and some final weighting has to be done. Most importantly, the phase of policy formulation involves assessing the effects of different strategies. Changing scenarios might give different results. The unexpected growth of one economic sector might influence the effectiveness of a certain policy. The quality of policy-making will be improved if some kind of decision support environment is offered, in which the effect of decisions in varying circumstances is calculated. The use of GIS in planning policy therefore rests to a large part in the integration of GIS concepts with DSS (decision support systems) concepts (Fedra et al. 1987, Ten Velden and Scholten 1988, Nijkamp and De Jong 1988). At present this integration is still in the experimental phase. The policy-making process in practice is still non-automated.

5.3   Present geo-information systems

At the present time, various GIS are being used at the RPD.

RUDAP

RUDAP, a system for spatial data processing, is by far the largest GIS application at the RPD. It contains a database of about 3000 variables on demography, land use, economy, infrastructure and migration. The example output of population density shown in Plate 5.1 indicates that the basic spatial unit is the municipality. The application allows for selection, classification, manipulation and aggregation of the variables in a flexible way and time series analysis is possible since the system includes data on yearly basis for the period 1970 to 1989. A transformation module caters for the changes in municipality boundaries in order to guarantee compatibility of data through time.

LBV

LBV is primarily a conversion system, which encompasses all addresses in the Netherlands, with identification of higher spatial units like neighbourhoods and municipalities. The LBV system is also used to aggregate data from a number of address databases to the level of grids (500*500m) (Figure 5.2). Address databases are available on recreation facilities, commercial services and public facilities.

BARS

BARS is a vector-oriented information system with data on about 40 categories of land use. Spatial entities from BARS are linked with attribute information on a number of subjects, such as traffic intensity, ownership and status of designated areas or characteristics of business districts. Furthermore, topographic elements are used as background information for a variety of applications. The data cover the whole of the Netherlands at a scale of 1:25,000.

LKN and CKN

LKN, used for landscape ecological mapping of the Netherlands is a conglomerate in itself. It has been constructed on a 1km by 1km gridcell basis and consists of a variety of ecological datasets including information on soils, landscape characteristics, flora, fauna, ecotope types, groundwater relations (Plate 5.2) and so on. The data for this information system are still being collected from a huge number of different data sources and with a range of conversion algorithms. Between 1989 and 1992, the database will be completed to provide coverage of the whole of the Netherlands since only the Randstad is fully incorporated now (Huzen and van der Schuit 1989).

CKN is used for cultural and historical mapping of the Netherlands. It is an information system used to support statements on the spacing or distribution, scarcity, interaction and change of historical landscape phenomena at a nationwide scale. Data has been collected in a raster of 2km by 2km squares on the scale of 1:25,000 for two years, 1840 and 1980. The inventory involves information on the parcelling of lots, build-up area patterns, infrastructure, parcel-edge vegetation, the phenomenona of peat turning, and the time period of the first cultural landscape design or construction (Profijt and Bakermans 1988).

Figure 5.2: Example of geocoding address data with the help of the LBV information system

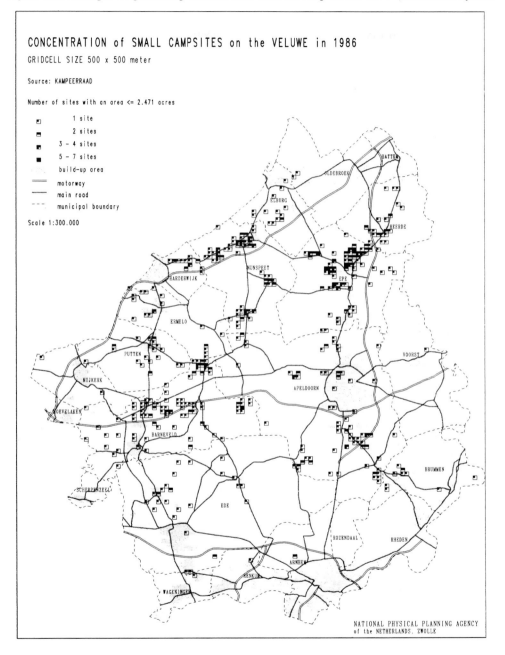

SvhL

The SvhL information system concerns the visible aspects or physiognomy of the Dutch landscape which are defined as the system of size relationships in the landscape. These are to be found in the pattern of both verticalborders (trees, dykes, buildings) and horizontal borders (plot edges, ditches, transitions in landuse). The vertical borders regulate the formation of spaces. The size of the spaces are termed 'breadth of view'. Within these spaces point or 'spacefilling elements' can occur, such as farms, which limit the view but do not form new spaces. Finally, there are horizontal borders, such as, ditches, which have a scale decreasing effect. The data has been collected in gridcells of 2km by 2km. The source material originated from the analogue 1:25,000 topographic maps.

LGN

The RPD is participating in a project that will result at the end of 1989 in a land cover database based on Landsat TM files. 16 land cover categories are distinguished, with an emphasis on types of rural land cover. The resolution is 30m by 30m and the database covers the whole of the Netherlands.

5.4   Resources and efforts

The development of these GIS has cost a lot of time, money and effort. The Information Systems Department of the RPD has been involved in this work since the early seventies. In those early days information was gathered and processed on paper maps and transparent overlay sheets. Since then geographic information processing has evolved through a digital CAD system unto a full GIS system within the ARC/INFO environment. Like every other computerized information system, GIS are moulded from a blend of hardware, software, data and orgware (Nijkamp and de Jong 1987). These four elements are now briefly described.

Orgware

In 1970, a seperate organizational unit consisting of three individuals was created and entrusted the task of gathering basic information of relevance to physical planning. Today the Information Systems Department consists of four sections with a staff of some 27 persons. The research and development section is responsible for information analysis (data scouting, gathering and conversion); database development; and application design and development. The system management section is responsible for the management of central hardware and system software, (including basic software packages); and for the selection and benchmarking of hardware and software including quality control. The geographical database section is responsible for geographical database management; development of geo-(carto)graphical datasets; and data and map delivery. Finally, there is the internal management section, responsible for all services that keep an organisation running like archiving, library, mail delivery, and canteen service.

One of the policies of the RPD is to collect and to use (convert, link, geocode) data from other data collection organizations. An extensive network of contacts with such organizations is maintained to ensure flexibility and ready answers to differing user demands. Most of the application development is contracted out to external software companies under project management of the research and development section. Database development is done in-house. Although the department's internal functions are primary, an increasing demand is placed upon its resources from external organisations, both from within the government and from outside. The delivery of data sets and maps to answer such demands is a task of the geographical database section.

Data

Four different categories of data sets are kept at the information systems department:

(i) Administrative data: a three-dimensional database of some 3000 variables for 800 municipalities and 20 years (see RUDAP). Source data are bought from the Central Statistical Office.

(ii) Grid data: based on 0.5km by 0.5km, 1km by 1km or 2km by 2km gridcells. Some 7 different data sets that cover the whole of the Netherlands are kept, mostly in the field of the natural sciences (emission data, land cover data, ecological data). These datasets are compiled by external research agencies with expertise in the field of interest.

(iii) Vector data of two types:
   (a) basic spatial structure data: topographical and functional landuse data digitized in vector mode on a 1:25,000 scale. This database is developed at the RPD and based on material from the Central Statistical Office and some 20 other organizations.
   (b) other vector datasets on scales of 1:25,0000 to 1:1,000,000 on a range of subjects, especially physical planning designated areas.

(iv) Address data: several data sets are developed that consist of address data with attribute information primarily on the subject of services. These datasets can be geocoded to a gridcell of 0.5km by 0.5km or a neighbourhood with the LBV system, a geocoding system developed at the agency that consists of all 6 million addresses in the Netherlands and some locational indicators. Today these differing datasets are in the process of being integrated into a central geographical database that will enable the linked manipulation of all the different spatial objects and their attributes.

Hardware

The main system consists of two Primes (9955 and 2755), one in The Hague and one in Zwolle, with a number of terminals (graphic and alphanumeric) personal computers attached, digitizers, plotters and printers. Besides a Prime 6350 serves the emission registration project, and two Sun/Unix computers and a 386 are used for database development.

Software

Operating systems are Primos, Unix and MS-DOS, although the trend is towards a complete conversion to Unix, at least in non-administrative automation. At the present time, the DBMS is Info, while the central database is being developed in Oracle. In the near future, Oracle will be the main DBMS as it is available on a range of computers from micro to mainframe. Basic GIS software is ARC/INFO, while on PCs, some use is made of packages like Atlas*Graphics and MapMaster. The purchase of a raster GIS package is being seriously considered. SPSS is used as the main analytical software tool (for statistical purposes). Some spreadsheets are used on PCs. Main fourth generation programming languages are AML (in conjunction with ARC/INFO), Info, Cwic (a fortran generator), dBASEIII and SQL with its extensions. AML and SQL will be the tools for the future. Third generation programming languages are FORT77 and C, with an increasing emphasis on the latter, especially because it is very suitable for interfacing. Graphical standards are PLOT10, GKS and CGI.

Costs

The wealth of instruments necessary to compose a stable GIS environment involves a lot of costs. The RPD numbers some 280 employees of which about 10% (27 persons) are employed in the Information Systems Department. Operating costs of the department are Dfl 3.4-4 million per year, composed of office costs and personnel (2,700,000), data acquisition (200,000), and hard/software maintenance (800,000). Some data sets like BARS are developed with internal support. Total development costs amount to some Dfl 8,000,000. Trends in geographical information processing involve data integration, the growing orientation of GIS away from experts and specialists towards the user (Swaan Arrons and van Lith 1984) and interactive decision support (Van der Meulen 1988).

5.5   Developing a central geographical database

On 1 January, 1989, a project to develop a 'central geographic database' at the RPD made a formal start (De Jong 1989b). Detailing of the functional design and software prototyping were the main activities in the first half year of the project. In this section, the basic ideas behind the concept are highlighted, the functionality is described and some comments on the project are made.

The concept of data integration is the fundamental characteristic of a central database. All datasets will be structured to a common model. All key items in this data model ensure accessibility on a variety of entries, such as period, area, subject. Central GIS applications will run directly on this database, while local applications will either run on selections from this database, or will develop a standalone database constructed along the lines of the central data model.

User orientation means decentralization of geographic information processing activities. The central database might be used in a distributive way. Parts of the database might be copied onto PCs or standalone workstations for further analysis. The central database is accessible from every computer, and facilities are realized to enable the user to retrieve his own selections without expert interference. This requires a graphical user-interface. For example, (iterative) retrieval on area, time or subject entries in an abitrary sequence might be facilitated by graphical images: a digital map, time scale or icons. The clipping of areas in a digital map does not mean clipping the database immediately. It is saved as one of the selection criteria, along with the time series and the number of subjects. This way of retrieval results in a user database that might be combined with specific user or project data, be it local or centralized.

To enable a quick and flexible method of manipulating data sets, attribute information is disconnected from topological information (not from spatial objects!). The central database (Figure 5.3) consists of three parts; the thematic database (all attribute information), the topological or situational database (all topological information), and the presentation database (all data necessary for the layout and visualization of products). A relational approach to data handling is chosen. The spatial object (SO), which might be a line, point or polygon element, constitutes the central connection between topology and attribute. By developing a central database in this way, the database management or analysis tools can be chosen freely. At present ARC/INFO is used as the main GIS tool. In this context, ARC/INFO does not structure the attribute information anymore, but only the topological information and an identifier of the SO. Other tools might be used concomitantly, like SPSS, SPANS, MAP, ERDASS or custom-built analysis software. In the analysis itself, topology and attribute information is often combined but sometimes seperated by the user. This approach speeds up processing and ensures conceptual clarity and elegance as well as organizational and management benefits.

Figure 5.3: Conceptual framework of the geographic database

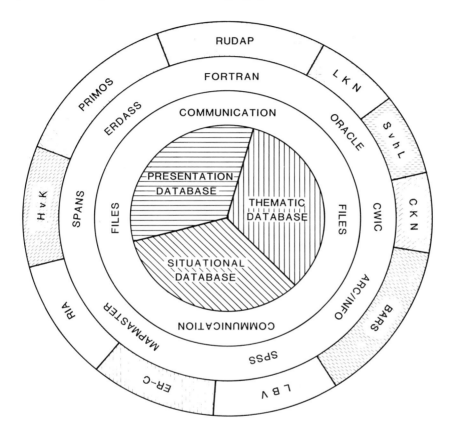

Another main concept of the central geographic database is a process orientation. This means that output is not defined before the process, by process parameters. In the presentation database, all parameters (for symbols, colours, shading for maps, for directing business graphics, tables etc.) are stored to ensure proper output. Output is composed on the basis of the selection or classification, together with the kind and type of data and the subject. Of course, user directions might interfere, but they are not necessary to produce correct output. A cartographic expert system is in development to perform this task. Descriptive information on the data stored in the database is an essential part of the user-interface. Knowledge about quality of data, source, storage (tape, disk) availability, relation with other data etc., is necessary for an adequate use of the database. It is possible to search through the meta data in various ways (hierarchical, semantic, keywords) to get an idea of which data are available on the subject to be studied.

By supplying meta-information, a graphical user-interface and the facility of user-defined databases, advanced user support facilities are created. Along with these issues, the issue of artificial intelligence in relation to database use is considered. The main idea is to use meta-information on data quality during the process of data manipulation to support and guide the user on the correct use of data and interpretation of results. The concept of certainty factors is drawn from the theory of fuzzy sets (see Klir and Folger (1988) on fuzzy sets and information; Drummond (1988) on an application of this theory in geographic data handling). By estimating the certainty factor of data, manipulation techniques and geographic models, it is possible to add a quality index to a product. Besides, in a 'what if..?' situation, meta-information might be combined with rules to predict quality. Such information might be used to decide whether or not to go on with a certain kind of analysis. It can be used to help proper interpretation of output and to show ways to improve datasets or geographic models.

Figure 5.3 shows how the database is surrounded by a shell of facilities for retrieval, manipulation and presentation. The applications all work with the shell of facilities. The processes that play a role in the use of the central database, and the way they use the several sub-databases are depicted in Figure 5.4.

Figure 5.4: Processes in the GIS database

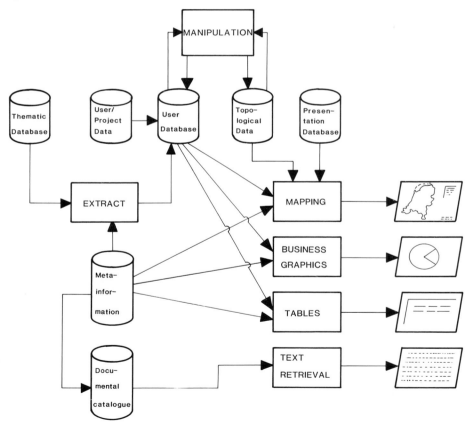

Thematic database

Within the thematic database, four storage sections can be recognized. Firstly, authorisation information is stored for each user by username and password identification. Each user will be able to work on one or more projects. Each project will be identified by a project name and number. Descriptive information on the project will be stored too. The actual entry will be dependent on the combination of user and project authorisation. Each project may use a selection (view) of Spatial Object-Types (SOT) stored in a project-SOT relation. A project might also be limited in the subjects for which information may be retrieved. This is stored in a project-subject-relation. Secondly, spatial information and administrative relations between spatial objects are stored for each spatial object. Each spatial object will have a name, belong to a spatial object type and exist for a certain period. A spatial object may be embedded in another spatial object or consist of certain spatial objects. This relational information might be extracted from the topological database too, but is stored here for efficiency purposes. Administrative relations between spatial objects are used very frequently in spatial analyses for physical planning. Thirdly, descriptive information about spatial object-types, variables, sources, and periods are also in the thematic database together with textual information and appropriate formats. Variables may be grouped into subjects. Which variable belongs to which subject is described in a subject-variable-relation. Subjects might be grouped themselves. Variables may be of a certain type and in the case of class variables, the class steps are also stored in the database. When the database also caters for processed data, the algorithm to process the data to derive the result may also be stored in the database. The fourth section involves the actual thematic data which might be linear, class, flow, flow-in-classes or transition type.

The described logical structure of the database will be (and partly is) implemented in ORACLE (thematic database) and ARC/INFO (topological and presentation database) in a UNIX environment. ORACLE is only used as a template for the datastructure. The actual thematic data themselves will be stored as flat ASCII files due to the volume of data that can be held in ASCII compared with ORACLE format, the ease of management and updating, and the simple structure of communication files between the central database and applications. By cutting down the amount of stored ARC/INFO data to that of pure topological data, the volume of the database is kept within manageable proportions.

Topological database

Geographic data can be defined as attribute data which are related to spatial objects in geographical space. These spatial objects can be administrative units, grid cells, physical units, point locations or any other type of spatial unit. The locations of SOTs and their spatial description (topology) are stored in the topological database. They are defined in terms of points, line segments and polygons of ARC/INFO. In the topological database the facilities of the Librarian sub-system of ARC/INFO are incorporated, so that it is possible to define activity space as a subdivision of equal or unequal tiles. For the Netherlands, the RPD has chosen a subdivision independent of time, administrative boundaries and physical units: the mapsheet subdivision of the Topographic Service of the Netherlands. The tile structure enables easy management of topological data. It speeds up access to the data and interactive retrieval by users. Three map libraries has been defined so far: scales 1 : 25,000 up to 1 : 50,000; scales 1 : 250,000 up to 1 : 500,000; and scales 1 :1,000,000 up to 1 : 1,500,000. The tile divisions of the smaller scale libraries are an aggregation of the basic division. The facilities of the topological database will support the use of the libraries in a hierarchical way. For the raster-based SOTs a special library will be created. Because ARC/INFO is vector-oriented, the storage of raster data is based on polygons or on the centre points of each grid cell. Every data layer represents a SOT at a particular moment in time. Administrative spatial objects in particular undergo boundary changes in time. The storage of all these changes in seperate and complete layers requires an enormous amount of external memory. In physical planning, where monitoring of phenomena is essential, the volume of data will keep on growing. To save on memory the topological database will dispose of facilities to store only the changes of SOTs as seperate layers and to construct (on user command) a new SOT of the desired moment in time by calculating changes backwards in time. The most recent layer of every SOT always contains all spatial objects of that type.

Figure 5.5: Choice processes in the cartographic expert system

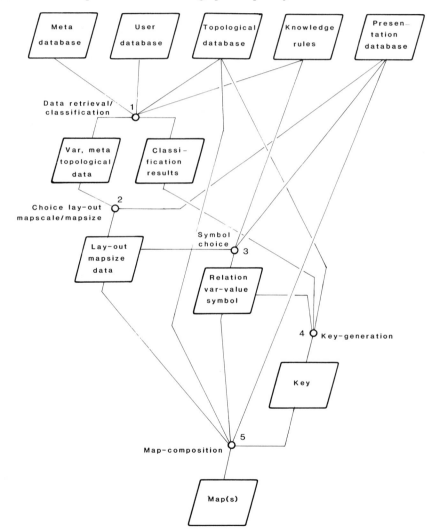

Presentation database

One part of the central database is specialized in storing the parameters necessary for the presentation of data retrieved from the thematic and topological sub-databases. Presentation can take the form of maps, business graphics, tables or texts. In the process of composing a map within the database at least five phases can be distinguished (Figure 5.5). The first phase involves data retrieval from the user database and the topological database and the (re)classification of variables. Secondly, it seems necessary to choose a map scale or map size and layout. The smallest spatial unit in the map, the number of data classes and technical constraints influence the decision on the map scale. Within the extent of the map sheet and the scaled topological parameters, map layout types are generated and displayed. The map will be composed according to the chosen layout type. The user might interfere and change the layout in the map composition process if he wants to.

The next two phases involve choice of map symbols and key generation. The selected (classified) variables are visualized by point, line or area symbols. Sometimes there is a direct relation between a variable and one symbol or a range of symbols. In most cases there is more than one possibility. In the presentation database are stored the definition of each map symbol (the colour, size, angle, type (marker, line or shade), number of definition layers) and the visual properties of the symbols. Using the meta-information about the variables (especially the measurement level and the subject code) and some cartographic knowledge rules, the range of possible symbols is reduced (Figure 5.5). Finally, the user chooses one of the displayed possibilities (if there is more than one left by the expert system). In the final phase, the map will be composed using the selected layout type and symbols. Parts of the title information, key texts, and map edge information (sources, date, etc.) can be generated automatically with the aid of the meta-information database. Topographic background information (from the topological database) can be added to the map. For certain subjects, the type of map background information is stored in the database. The incorporation of expert knowledge addressed at the symbol choice process needs a set of complex data structures, facilities to store rules and facilities for knowledge interpretation. The definition of symbol ranges, the visual properties of map symbols, the measurement level and the organisation of variables are stored in the relational database of ORACLE. At the end of this year a prototype of the presentation database with the cartographic expert system will be ready for operation.

5.6   The future

By a systematic approach to illustrate the activities of the Agency in (potential) relation to GIS, the main requirements of a spatial database suited to the goals of the organisation are discerned. So far the geographic database development project concentrated on specific processes described above (data model, query, mapping) to meet part of these requirements. Other processes that play a role in the use of a geographic database in the Agency have to be supported too. In the coming years, some of the main development goals in this project (or in its offspring) will be intelligent user support. The cartographic expert system as mentioned above might be extended from the support of proper map layout and graphic symbol choice to include the notions of map type and map purpose and the influence these notions have on map outlook. Another main theme will be the introduction of knowledge-based spatial data handling. One of the main drawbacks of traditional GIS is the 'hard' nature of its operations on data. In physical planning, decisions are often based on more 'soft' notions of reality. The development of a set of predictive rules that give insight into the level of uncertainty of manipulation results is on our research agenda.

References

Bertin, J. (1974) Graphische Semiologie, Berlin

Drummond, J. (1988) Fuzzy sub-set theory applied to environmental planning in GIS, Proceedings of the Eurocarto Seven Conference on Environmental Applications of Digital Mapping, Enschede, (September 20-22)

Fedra, K., Zhenxi Li, Zhuongtuo Wang and Chun Zhao (1987) Expert Systems for Integrated Development, A Case Study of Shanxi Province, The People's Republic of China, IIASA Report, Laxenburg

Huzen, L. and van der Schuit, J.H.R. (1989) An overview of environment related information systems at the National Physical Planning Agency, Paper presented at the Seminar on Environmental Mapping, ITC, Enschede, (April 25-27)

Jong, W.M. de (1989a) Uncertainties in data quality and the use of GIS for planning purposes, in Proceedings of the UDM Symposium, 1989, Lisboa, pp. 171-186

Jong, W.M. de (1989b) Development of a large national geographic database in the Netherlands, Paper presented to the 1989 SORSA Colloquium, Maryland, USA, (March 29-31)

Klir, G.J. and Folger, T.A. (1988) Fuzzy Sets, Uncertainty, and Information, Englewood Cliffs, New Jersey

Meulen, G.G. van der (ed) (1988) Informatica en Ondersteuning van Ruimtelijke Besluitvorming, Eindhoven

Nijkamp, P. and Jong, W.M. de (1987) Training needs in information systems, Development Dialogue, 8(1), Spring

Nijkamp, P. and Jong, W.M. de (1988) Informatica en ruimtelijk beleid, in Meulen, G.G. van der (ed) Informatica en Ondersteuning van Ruimtelijke Besluitvorming, Eindhoven

Profijt, I.R. and Bakermans, M.M.G.J. (1988) Cultuurhistorische Kartering Nederland, in Geografisch Informatiesysteem, Report No. 1954, Stiboka, Wageningen

Schuit, J.H.R. van der (1989) Cartographic aspects of a central geographic database, Paper presented at the ICA Conference, Budapest, Hungary, (16-25 August)

Schans, R. van der (1988) De geografische gegevensstroom in kaart, Geodesia, 88(4), 552-554

Swaan Arrons, H. de and Lith, P. van (1984) Expert Systemen, Den Haag

Velden, H. ten and Scholten, H.J. (1988) De betekenis van decision support systems en expert systems in de ruimtelijke planning, in Meulen, G.G. (ed) Informatica en Ondersteuning van Ruimtelijke Besluitvorming, Eindhoven

Wouter M. de Jong
Rijksplanologische Dienst
Afdeling Informatievoorziening
Blijmarkt 20
Postbus 502
8000 AM Zwolle
The Netherlands

PART III
URBAN PLANNING APPLICATIONS

6   INFORMATION MANAGEMENT WITHIN THE PLANNING PROCESS

Frank le Clercq

6.1   Introduction

Planning and information are closely linked. Information is needed to assist decision making in planning, and the monitoring of urban development may result in new planning actions. It could be argued that in the preliminary stages of planning, when discussion is focused on plan making rather than implementation, planning is almost synonymous with information organization. At this stage, information management and planning process management coincide. The present chapter is based on this conjunction. Once the operations of planning processes are known, information provision can be managed accordingly. The chapter will show how planning processes can be disaggregated into elements which occur in several types of oprerational process in planning. These elements may serve to manage processes and to formulate planning information requirements.

The approach of this chapter differs from methods that order information according to substantive themes or to types of plan. Substantive themes, such as traffic, population or employment, are a basis for planning information because they represent knowledge and statistics. This view of information requirements is supply oriented because the availability of data and knowledge directs the structure of information provision.

Another approach is to structure information needs according to plan types. The statutory plans, like local plans and structure plans, are the criteria which determine information needs. This view is demand (planning need) oriented, but it does not imply a uniform need for information for each plan type. For instance, a structure plan can be set up for a variety of reasons. It might aim to offer an integrated analysis of related problems in a city or to open new perspectives through an inspired view on the future of the city. Its planning content may range from being the legal basis for local plans to the provision of a complete program of projects to be developed. The information needs will vary accordingly.

It is convenient to regard information requirements as resulting from behaviour in planning practise. For this purpose the planning process will be decomposed into several elements that must be considered. Thus, the discussion begins with a characterization of planning practise using Amsterdam's waterfront plans as an illustration. The generalization of plan processes reveals the common elements of process management and information provision. These are summarized because a more theoretical explanation is given elsewhere (Le Clercq 1990). The information requirements associated with the planning elements are then formulated, and special emphasis is given to the information that geographical information systems can supply to planning processes.

6.2   Planning process management

At the present time, the vocabulary of planners is rapidly changing. Such terms as city marketing, public-private partnerships and target groups point to a change in planning activities and perspectives. Planning is becoming more diverse with respect to its very modes of

H. J. Scholten and J. C. H. Stillwell (eds.),
Geographical Information Systems for Urban and Regional Planning, 59–68.

operation. In the past, comprehensive planning was the main planning method; all aspects of a plan were brought together in a balanced and coherent way. Today 'selectivity' is the key word, with respect to sites to be developed, issues to be addressed in plans and clients to be served. The objectives that may influence the selection include creating leverage, meeting social weaknesses, exploiting area-inherent strengths and seizing new economic opportunities.

These changes can be briefly characterized by saying that approaches in planning aim at urban development and apply business-like methods. Planners are not only responsible for the preparation of proper decisions and for planning control, but also for marketing plans, identifying strategic issues in the implementation of plans and creating portfolios (packages) of projects in which financial advantages and limitations are balanced. The changes form a challenge in planning process management. A great variety of methods, techniques, approaches and operational procedures have been introduced to help the planner to achieve the desired aims. This variety of planning actions requires an effective management of the planning process. The right issues should be addressed by the most efficient and appropriate planning approaches.

To study the various planning approaches, a schematic framework has been developed which combines Simon's descriptive rational model of administrative behaviour with the strategic management model and observations about management behaviour (Le Clercq 1990). The rational model of administrative behaviour was introduced in 1946 (Simon 1967). In Simon's opinion, administrative procedures aim at rational decision making; that is the selection of the most appropriate option out of a set of alternatives designed to solve a well defined problem. This model can be further rationalized with regard to the organization of public sector planning agencies, legal planning procedures and planning tasks for the various tiers of government (Faludi 1973). In planning, the rational model has mainly been interpreted as a normative model of what public sector decision making should be about. However, the model can also be seen as a description of the ways in which problems can be dealt with. Simon himself has developed this line of thinking in his work on artificial intelligence (Newell and Simon 1972). The view holds for the planning field too. The rational model can be seen as a model of the way in which planning-type problems are dealt with. The model points to elements about which information is needed for progress in a working process: the precise problem; alternative solutions to the problem; and decision making criteria to assess the alternatives.

Studies of management behaviour show that approaches other than the rational model are also conceivable. Such models introduce a number of operational procedures or heuristics that help to manage the planning process and to acquire information. These models include an element that is not part of the rational model, namely opportunities for planning actions to achieve results. This element is particularly important in the strategic management model (Bryson and Roering 1987). An example from planning practise in Amsterdam will illustrate how all the elements that have been mentioned figure in a complex planning process. Subsequently, it will be shown that planning activities can be generalized into a schematic framework consisting of the elements identified.

6.3 An example: the start of a development plan

The IJ-axis development plan takes its name from the River IJ, where the port of Amsterdam has always been situated. Plans exist for the redevelopment of the former harbour areas into a top-of-the-bill residential, leisure and office complex on the waterfront. The status of Amsterdam's harbour areas on the River IJ is that of a reconstruction area: new functions will be introduced and new physical developments will take place. The City of Amsterdam has a standard procedure for dealing with physical developments on land owned by the City. The City Council decides, in a multi-step procedure, details about the development programme, investments and the lease of land. This is a completely controlled planning process. It does not, however, guarantee that developments will actually take place. These rely on initiatives taken by private developers. To incorporate these market opportunities in plan making, the planning process must also deal with market needs. Thus, actual planning actions should include much more than the design of physical plans and the preparation of associated decision making. Analysis of the initial stages of the IJ-axis plans shows that the working process is a direction-seeking effort rather than a straightforward development procedure.

The problems

For a long time the city has faced a number of problems which are associated with the IJ-area; or problems that are thought to be solvable by means of new constructions on the waterfront:

(i)   the harbour areas are deteriorating because more modern basins have been constructed to the west of the city;
(ii) offices are leaving the city centre to locate in suburban centres whilst demand for office floorspace might be met in the former harbour areas close to the city centre;
(iii) access to the city centre is poor although a new arterial along the IJ-axis may solve this problem;
(iv) the City authorities are constantly looking for central city sites to construct houses to stop population decrease. The former harbour areas offer an opportunity to accommodate new housing too.

Initial impetus for development actions

The problems mentioned above had resulted virtually no concrete planning initiatives in the former harbour areas. Plan making had only begun in connection with a residential area designated in the most eastern part of the area. City planners only became aware of the quality of the site when the Dutch Railways reconstructed the interior of the Central Station. This station is situated on a very prominent spot on the IJ-axis, close to the city's historic centre and on the water's edge. At the same time, the waterfront movement in the U.S.A. had shown very promising achievements. Consequently, it was decided to redevelop the waterfront close to the Central Station as a public resort area. To that end, access from the city centre to the waterfront has to be gradually improved.

Creating an image

A contest for town planners was organized to gather ideas for waterfront designs and solutions for the linkages between the existing historic city and the waterfront. A synthesis was made of the best ideas.

Looking for opportunities for development

This synthesis was used to market the ideas to real estate investors and developers. Under what conditions would they be willing to invest in the area? One of their requirements was better access to the development area for traffic coming from the urban ring road. The national government was drawn into the process in two ways. Firstly, through the budgets that were established for the reconstruction of large scale housing projects on inner city sites. Secondly, the national goverment, in its Fourth Policy Report on Physical Planning, announced the need for a top office location in the Netherlands, which might be in Amsterdam.

Developing project options

In the next stage of planning, the designated area was subdivided into several project areas, for which investors and developers were asked to submit proposals for development which would meet pre-formulated functional programmes.

Recasting image: policy formulation

To define the functional programs and to meet developers requirements, the image of the area was updated by reviewing the previous plans and formulating final development policies. The process of going forward and backward through the cycle of process elements (Figure 6.1) and of selecting successful projects is now underway.

Figure 6.1: Successive steps in the IJ-axis development plan process

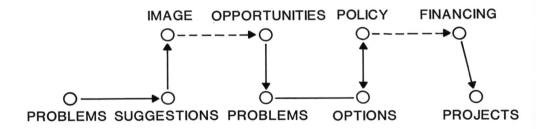

6.4   A synthesis of planning processes

The five elements that may compose planning cycles, have been brought together in Figure 6.2. The elements are consist of:

(i)   the problems to be met;
(ii)   the opportunities and threats pertaining to the planning issue;
(iii) the images of the situation, usually expressed by means of policy statements and sketch designs;
(iv)   the options for decision making, mostly at the project level; and
(v)   the decision making process after considering the impacts on all the parties involved.

A number of paths can be traced for a single project, a group of projects or the plan-making for an entire area. This is evident if Figure 6.2 is considered as representing one single cycle in a dynamic process that goes through several cycles. There are reasons to consider some process elements several times, as the IJ-axis example illustrates. Firstly, when initial ideas had been formed, a second cycle was started, looking for development opportunities. Secondly, new problems raised by likely developers, started new policy formulation activities. Thirdly, in order to implement part of the policies, a new cycle was started at the project level.

Several types of process can be described from the elements of Figure 6.2. These are the commonly known planning processes. They include courses of action like: the rational planning process (problem-options-decision); the policy development process (policies-selected projects-options-decisions); and the strategic management process (opportunities-problems-policies/missions-options-decisions).

Process management can be considered as selecting the right course of action expressed by the various planning elements. It has been argued elsewhere (Le Clercq 1990) that a proper criterion to select action paths is the reduction of uncertainty by considering the uncertainty associated with the various planning elements. As uncertainty is also a measure for the information content, the management of actions in the pre-decision stages of planning agrees with a proper supply and selection of information included in the planning elements.

Figure 6.2: The elements of a single planning process cycle

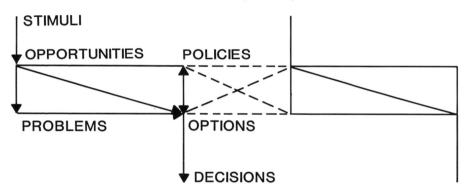

6.5   Information requirements

The conclusion of the previous section implies that the planning elements also provide the basis for identifying information needs. To elaborate this view, the way information is processed in planning is demonstrated initially. Then some further steps are taken with respect to the elements: the kind of information to be associated with the elements has to be indicated and operational procedures and heuristics to gather information must be known to identify the right sources of information. We shall only deal with the processing of information and the type of information related to the various planning elements. Operational procedures have been discribed elsewhere (Le Clercq 1990).

Information processing in planning

Planning usually starts from knowledge about attributes of the phenomenon. In the case of the IJ-axis project, these might include the likely demand for office space, the lack of housing in Amsterdam, the total transport flow to the central city and so on. With the help of this information a feasible design is produced. The information concerns attributes. In planning practise this information is both of a descriptive nature (about the characteristics of the planning issue) and of a normative nature (goals to be achieved by the plan). The design itself may be called a holistic representation of the various attributes. It can be called an entity which offers, by the very combination of values of the attributes, a new synergetic perspective on the attributes.

Thus, there are two important ways of dealing with information associated with planning. Firstly, by means of attributes of the phenomenon and secondly, by means of holistic representations, such as images and designs of the planning problem and the objects to be realized. The attributes may be inputs in the planning process or impacts to be tested at the end of the process. The representations concern the planning entities that one is dealing with, which are mostly 'designs' in either stage of development. Attributes appeal to researchers. The description of an entity by attributes allows its representation in a databank and subsequent deterministic analysis. Decision makers and designers mostly prefer to think in terms of physical entities, like buildings and parcels of land. Storing the images of these entities in a database is more difficult. Geographical information systems can bridge both ways of processing, because they can store entities and attributes. Figure 6.3 illustrates a scheme representing the two steps through which a process may pass with respect to attributes and entities.

Figure 6.3: The planning process as attributes and designs

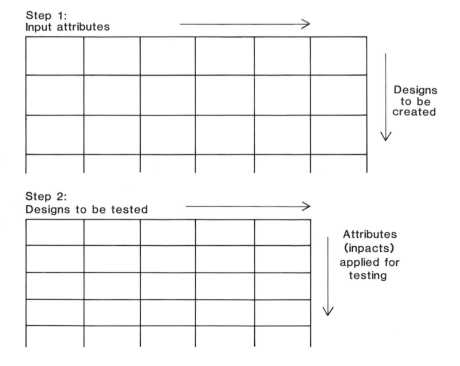

The first step is known as the design process. In this step, attributes are given at the outset (from previous analysis of the phenomena) and the number of feasible designs is unknown. The second step is known in planning as the evaluation or assessment stage. In this second step, the designs are known, but they have to be tested for a number of attributes. The relevance of each of the attributes still has to be proved. This rather abstract expression of planning information in terms of attributes and designs or holistic representations provides the capability of testing what information every planning element may contribute to the planning process.

Decisions

Options are needed for decision making and attributes for testing the options as to their decision making value. In physical planning options will concern physical entities, although the decision itself may only be about the allocation of money. So, both the holistic and the attributional way of processing information are relevant.

Opportunities

Three types of opportunities or threats (negative opportunities) exist in planning:

(i)   trends or exogenous developments, like population changes, shifts in the economy or changes in public awareness of the seriousness of environmental issues. The information concerns attributes (inputs);
(ii)   opportunities to realize projects, like a subsidy program for urban renewal. The information content is of an attributive nature, because most of the opportunities will only be conditions to further project development;
(iii)   real proposals for projects from investors or companies. These will be added to the planning options at the project level. The associated information concerns entities.

Problems

Problems can be distinguished by their origin. They can stem from an opportunity or threat (external influences) from internal weaknesses or malfunctioning or from development requirements, which are problems resulting from the further elaboration of an accepted policy. Most problems will concern planning attributes, except some of the development requirements that ask for designs.

Policies

Policies are value statements about goals to be achieved and constraints to be considered. The information mostly concerns attributes, expressing goal variables such as necessary population development and environmental qualities to be achieved, or constraint factors like budgets that have to be met. Policies may originate from opinions one has about likely solutions or from preferences for particular options. In the latter case, policies might be expressed as entities one would like to have developed.

Options

Solutions or options for decision can have their origin in policies or problems. Options derived from policies can be seen as examples that explain the policies one would like to have developed. For instance, some of the designs for the IJ-axis projects only served to show to developers how projects might meet the set of policy statements. Developers themselves can propose real projects as answers to well-defined planning problems. Both types of options or solutions are represented as entities and by the attributes that are relevant for decision making. Table 6.1 combines the information needs included in the elements.

Table 6.1: Information needs by planning element and processing characteristics

Planning element	Information characteristics	
	Attributes	Entities
Decisions	Impacts and resources	Options
Opportunities	Trends: - population - employment - economy - etc. Planning opportunities - investment programs - subsidies	Development initiatives: - buildings - land - activities
Problems	Analysis of trends; Strengths/weaknesses: - traffic problems - social issues - economic shifts Requirements to meet - see policies	Implementation programs: - projects by time and resources Designs for: - city renewal - buildings - city block - city extension
Policies	Goals: - population development - environment quality - accessibility - etc.	Pet projects
Options/Designs	Impacts: - traffic - environment - economy - external effects Planning data: - costs - land occupancy - density - etc.	Designs: - layout of parcel of land - building - transport network - activities Proposals for - constructions - activities

6.6 Geographical information systems in planning

Geographical Information Systems (GIS) are well known for their ability to store statistical data, for the analysis of trends and developments and for the presentation of this information geographically. The scheme of Table 6.1 however, suggests some further applications. If a GIS has to support planning processes it must also provide information about existing opportunities, problems still to be solved, policies to care about and options under consideration. It should at least serve as an easily accessible archive to planners, an archive organized according to the elements of a planning process. The information should be dynamic and must be continuously up-dated to the state of the planning actions.

A GIS can be more than this. It can be the software instrument by which to create a planning process support system that associates planning elements with each other. This feature can be accomplished because of the ability of a GIS to comprise geographical entities. These are entities that can be characterized, as in the real world, by cognitive descriptions such as office building, the design for a leisure facility, development proposals by company A, or public transport links/stations. The entities can be uniquely determined by a GIS and linked to each other by their geographical reference. Several attributes are related to the entities. These attributes can be associated with other planning elements that comprise attributes from a different planning perspective, like opportunities and goals. In this way it is possibls to test which projects achieve which goals. An example of a planning support system based on a GIS can illustrate the operations of such a system. The system should be built along the following lines.

(i)   The planning area should be well-defined. It can both be a restricted area, like the IJ-borders or a complete city if a structure plan is involved.
(ii) The geographical reference system can either be the existing topological structure (building blocks and network links) or be composed of the integration of the existing situation and reconstruction plans. For this purpose it should be very easy to introduce new building blocks and links in the system.
(iii) With each elementary topological unit or cluster of units, a description of the functions at the unit should be provided: house, office, road, vacant site, etc.
(iv) The description of the unit, though, should also be separated for the various planning elements in order to associate the entities of the various planning elements with each other: does an opportunity really fit with a plan option?
(v)   Several characteristics may be related to each of the descriptions. These can both be of quantitative nature, like the amount of floor space to be realized, or of qualitative nature, like the name of the land owner, the status of the proposal for development, or the kind of function at the unit. These characteristics can be stored in two types of systems, a database system oriented towards figures and a text reference system, like LOTUS-AGENDA (Manheim 1989). The addresses of both systems, however, should be the elementary topological units or clusters thereof.

Consequently the system can support several planning actions. It may work as follows:

(i)   If there is an opportunity, an immediate check can be undertaken of whether plan options exist that meet the opportunities and if the opportunity agrees with (location-oriented) policy statements.
(ii) In order to develop options the policy statements (attributes), development problems and opportunities can be associated with the area and serve as a programmatic starting point for the development of designs.

(iii) Strategic projects (plan options or decisions) can be appointed by selecting those projects meeting opportunities, trends and policies, as well as staying within resource constraints.

(iv) Policies can be determined by extracting from plans the arguments (attributes) that appear again and again to support decision making. These arguments point at value statements and resources that have to be considered.

6.7 Conclusion

This discussion has aimed to introduce a new approach to the analysis of information requirements in planning. This is done by associating information with the elements which constitute a planning process, such as the problem definition, opportunities, policy statements, options for decision making and decision-making impacts. The information requirements as such can be expressed by planning entities, like buildings, roads, parcels of land, and by planning attributes, like costs, floorspace, type of activities.

Entities in particular can be associated with units having a reference in a geographical information system, like a building to a block, or a road to a line segment. In this way, a GIS may serve as the basis for a planning support system, to link qualitative descriptions and characteristics of entities to their quantitative attributes. If this information is separately available for the various planning elements, a GIS can be applied to support the progress of planning: what are the relationships and associations among planning elements and which missing information should be acquired to complete the necessary insights in the planning elements?

The information about the status of planning elements is needed to manage the planning activities: what element should be worked on next: problem analysis; policy formulation; design manufacture; search for opportunities, decision making? Thus, information is needed for the successful management of planning processes.

References

Bryson, J.M. and Roering, W.D. (1987) Applying private sector strategic planning in the public sector, Journal of the American Planning Association, 53, 23-33

Le Clercq, F. (1990) Information supply to strategic planning, Environment and Planning B, Pion Ltd

Faludi, A. (1973) Planning Theory, Pergamon Press, Oxford

Manheim, M.L. (1989) Towards true executive support: managerial and theoretical perspectives, in Widemeyer G. (ed) DSS 89 - Proceedings of the Ninth International Conference on Decision Support Systems, The Institute of Management Sciences, Providence, R.I.

Newell, A and Simon, H.A. (1972) Human Problem Solving, Prentice-Hall, Englewood Cliffs

Simon, H.A. (1967) Administrative Behavior, (3rd edition), The Free Press, New York

Frank le Clercq
Dienst Ruimtelijke Ordening Amsterdam
Wibautstraat 3
1091 GH Amsterdam
The Netherlands

7 FIXED ASSET MANAGEMENT AND GEOGRAPHICAL INFORMATION SYSTEMS IN THE NETHERLANDS

Peter H.A. de Gouw

7.1 Introduction

Local authorities are required to perform a variety of functions to maintain or improve features of the physical environment which fall within their jurisdictions. The management of fixed assets, those primarily man-made features which exist below, on or above the earth's surface, is the responsibility of local governments in the Netherlands. It is one area of management where computers have not been employed on a large scale until relatively recently. Other areas have been automated for a long time and second and third generation applications are already being replaced by newer software. However, for fixed asset management, no application tools have been available.

Thus, the opportunity has arisen to implement state of the art computer technology in a situation which has had only limited exposure to computers. The challenge has been considerable one, and this chapter aims to expose some of the benefits and difficulties associated with the implemention and use of a GIS in this context. Some consideration is given firstly to the structure of local government and the extent to which computational facilities are now used by particular departments. Characteristics of a GIS custom-designed for medium-sized municipalities such as Amersfoort are then described, before the problems and advantages of the system are identified.

7.2 Local government functions

Local government areas in the Netherlands are usually divided into groups based on size. The most frequently used criterion for classification is that of population size, and typically the following five groups are identified:

Municipalities with a population of:
(i)   Less than 7,500 - Very small
(ii)   7,500 to 20,000 - Small
(iii) 20,000 to 50,000 - Medium
(iv)   50,000 to 120,000 - Large
(v)   More than 120,000 - Very large

However, in dealing with land management aspects, the actual size of the municipality (in acres) may become a more important determinant. In this chapter, we concentrate on the medium and large local government areas, since most of the small and very small organisations are not equipped to support a complete set of fixed asset management applications. The very large municipalities (in total less than 10) are in the process of developing fully customized applications to meet their requirements. However, it should be kept in mind that exceptions do occur. There are no technical reasons why municipalities have not been involved in using applications of GIS. It is either a matter of choice (for the very large) or the lack of organisation (for the very small).

H. J. Scholten and J. C. H. Stillwell (eds.),
Geographical Information Systems for Urban and Regional Planning, 69–76.
© 1990 *Kluwer Academic Publishers. Printed in the Netherlands.*

Every local government has a number of tasks it has to perform. If we group these tasks together we can distinquish four different types: inhabitant registration; financial management; social and welfare functions; and fixed asset management. Each of these can be elaborated in a little more detail. Inhabitant registration is a legal requirement, and recent legislation from the government describes in detail how information regarding each person should be handled. It also defines clear rules on how information should be passed to other interested organisations. The latter can be divided into three groups: other users within the same local government; users in different local governments; and authorised parties such as the Central Bureau of Statistics or the police. The most important legal issue regarding inhabitant registration states that this information must be regarded as original. This ensures that the correct information is available for use (name, address, etc. correctly spelt). As a result of this legislation, all users within the local government organisation, as well as the other parties mentioned, have to use this specific information set. This means, in effect, that any change of registration information, irrespective of the department concerned, should always involve data from the inhabitant registration file. Computers have been used to handle this information for some time, but a new generation of software has been developed in response to the new legislation, an extensive part of which deals with communication, since the law describes in detail how and what information should be passed on to other users of the data. The application is handled entirely by the single department responsible for registrations.

A number of years ago another law was passed that defined how the financial management of local government should be dealt with. This law does not just apply to one specific department but describes the rules and procedures for financial management across departments. Financial departments have used computers for a long time themselves. Newer and better software is adopted continuously and more and more tasks are executed using computers. Although the main activities concerning finance are handled by a financial department, other departments do use the financial systems too. Computers are being used increasingly for the management of social security by the department concerned, and here also, like registration, the total application is handled by one department.

Finally, there is the management of fixed assets which can involve a number of different departments, and which does consume a lot of resources. Some of the tasks included in this category are now handled by computers, but many remain uncomputerized. Whilst departments responsible for finance and social security are linked by computer network in order to perform these functions effectively, the exchange of information about fixed asset management between departments still tends to be a manual rather than an electronic activity.

7.3   Fixed asset management and spatial data

Fixed asset management covers a wide range of tasks and most information handled by a local authority has a spatial component. This component can be an address, a zip code, an area or neigbourhood, a land parcel, or a building, for example. Although no precise figure exists, the estimate is that between 60 and 85 percent of all information processed by a local government has a spatial component. Until recently, cadastral, thematic and topographic maps have been used to express spatial components, and one problem has always been that not everyone can use the same map. In effect, this has created an environment in which a large number of different maps are used, and although most of them originate from the same base map, this has complicated the general use of spatial data.

The number of departments using spatial information in any form depends on the size of the organisation and how departments are organised. Departments that are technically oriented (surveyor's, planning, roads, sewers, environment, green) will tend to use maps for different purposes than departments that are administratively oriented (tax, building and living, residence, legal affairs, cadastre). The surveyor's department is the main source of maps, with the capability of undertaking its own survey activities, map creation and maintenance. Based on survey maps and combined with new information, the planning department creates new maps that will propose changes. If accepted, these planning maps will either go back to the surveyor or be further altered by the design or the building department. The roads department will require surveyor's maps for maintenance and planning, and survey maps will also be used as the basis for information about the sewer and waste water system. Additional information will be included and once installed, will be very difficult to remeasure. Thus, much care must be taken to avoid information loss. In many parts of the older cities, where sewer sytems can be more than a hunderd years old, the exact location of a pipe is very hard to find in the first place. The availability of maps has become increasingly important in the tasks undertaken by the environmental department. Whether dealing with toxic waste, a leaking sewer or a company with permission to handle chemicals, location is an important element. Green departments are those responsible for looking after all the flowers, trees, grass, bushes, etc., in the municipality. The amount of detail collected in this context is often down to the level of keeping track of each tree or flower bed.

The departments concerned with administration will tend to use maps for different reasons. The taxation department requires housing maps to levy taxes, whereas the building and living department uses maps to monitor planning permits and undertake development control. Spatial information on ownership of land parcels is required by the cadastral department to prevent any prohibited resale of property in populated areas (i.e. as a means of preventing property speculation). The police and fire departments of municipalities also make use of both technical information, such as access routes or fire hydrant location, and administrative information, such as crime rates or fire hydrant capacities.

7.4   Computer applications

It is the development of modern GIS techniques that has brought about the development of computer applications in the various areas mentioned previously. The basic model begins with the construction of a central data set. This data set covers all information that is used by more than one department. This typically contains information on buildings, parcels of land, streets and addresses, persons and organisations, topographic elements and area boundaries. Each application would have its own specific data but would also use the central data set. This data is not limited to administrative information but will also cover geographic information. Benefits are derived from having all applications using the same central dataset. In this way, not only is duplication avoided, but a uniform definition of data and its use is maintained. However, by creating a modular framework, the municipality can start with a certain number of applications and add to these over time, the final goal being a totally integrated fixed asset application platform.

In order to build actual applications, a platform is required, which can be described as a data base that will hold both administrative and spatial data and manage that data for a large number of users. A real GIS is required to handle large amounts of administrative and spatial data simultaneously, and to integrate the two datasets in a multi-user environment. To actually accomplish this, the system has to be able to store all geographic data in a continous data base, no longer based on map sheets. The GIS should provide the user with methods for complex spatial analysis, specifically designed for an evironment that deals with a large amount of detail and which has the highest possible accuracy. A number of systems are currently being developed, one of which is the Avanced Relational Geographic Information System (ARGIS) from Unisys. Plate 7.1 illustrates the screen display of a small area chosen by the Amersfoort municipality to use as an example for the application of ARGIS. A variety of information (buildings, land parcels, roads, street furniture, etc) is presented.

ARGIS has a completely new design and has been built following a number of guidelines. Overall, the system is based on computer hardware and software standards. Thus, all the computer hardware is UNIX-based, and the network facilities are based on standards such as Ethernet TCP/IP (IEEE 802.3). The database is the ORACLE relational database, window handling is performed using X-Windows, standard exchange formats such as DUF (Dutch Data Exchange Format) are supported, and all software is written in the C language or in SQL (ORACLE's Query Language). Another of the guidelines used is that the system should be based on distributed processing. This involes some processes not being executed on the original database but on a temporary copy or extract. To make full use of the available computer technology, these processes are then executed on graphic workstations and do not affect the central computer system. A user friendly system is also required which does not limit use of the system to those with the necessary technical skills. Less skilled users who do not use the system all the time should still be able to work with the system. The user interface is created using menus and windows as well as help facilities.

Since GIS users cannot be limited to a single company or organisation type, a toolbox approach was selected which makes it possible to fully customise the system to the user's requirements. Through a number of modules, the system can be customised to client requirements which may vary between different symbol libraries to fully redesigned and translated menus or newly created menus, graphic functions and database structures. Data storage is no longer limited to a map sheet orientation. Storage is seamless or continuous, both for the administrative as well as the geographic data. Networks such as sewers, roads or utilities are not limited to a map sheet but can be designed, traced and analysed across the total area. This facility of continuity means that any area can be selected to be processed and database design is not restricted to one area.

In order to fully integrate geographic and administrative data, a special symbol language is developed allowing the user to control graphic element representation. This symbol language can be created for each department or user, allowing presentation to be customised. Any feature whether it is a point (symbol), line, polygon or text can be displayed according to user or data base criteria. For example, a tree can be represented by a selected symbol of chosen size in any colour, and displayed with administrative information as requested, under the control of the user and not limited by the original data capture. A surveyor's department will only mark its location with a single symbol. A green department may want to see a different symbol for the type or quality of a tree, as well as its unique identification number. Occasionally, a different size of symbol may be required for presentation. This is possible without making any changes to the data or the way that information about the tree is stored. Plate 7.2 is a detailed section of Amersfoort showing roads, buildings (including house numbers) and greenery (including identification numbers). Attribute information on particular buildings can be presented in a separate window, as indicated.

Since the whole system is designed for several departments and consequently a large number of users, an extensive security and content management system is incorporated. Storing and managing large multi-user data bases requires proper security access. However, in ARGIS, a further complication of security is the combined availability of administrative and geographic data. Access and update security can be introduced, down to the user level. An extensive locking mechanism is available to protect the data from double updates. Parts of the data can be updated from a remote location (workstation) while at any time allowing users to access the data for viewing purposes. In general this means that ARGIS is not a GIS application, but a tool to create that application. In order to provide local governments with a total application platform, Unisys The Netherlands creates a total set of applications for fixed asset management, all of them using ARGIS.

CLOVIS is the name of a concept for fixed asset management designed by CMG (Computer Management Group) and the Technical University in Delft. This concept is used as a basis for application design and creation by four vendors. This reduces the cost while still allowing vendors to distinguish their own offering, whether it involves a proprietary operating system or an open system. Differences are also enhanced by the software chosen to handle the spatial information, which may vary between a CAD system or real GIS software. The basic concept of CLOVIS is that a certain part of all the required data is available to most of the users. By defining this data, which is not limited to administrative data but includes geographic data, a standard for all users is created. By also defining the ownership and maintenance of the data, all users can benefit from better quality while spending less time on creating and maintaining the data. The benefits are therefore not just more and better information, but also uniform access to this information. It becomes possible to access data by using the spatial part in the system. This spatial part can be a neighbourhood, a district or any other that is available or can be created. It also allows data to be selected directly through the graphics, and presented to fulfill our requirements.

Every application that is created using the CLOVIS platform always uses information from the common data base, but also adds its own specific data. Most users will have access to one or more applications, but will also have access to the common data. This allows users to combine their own information with information used by other users or departments. A road maintenance application can view a road and ask the system for information on the road, but also on elements surrounding or below the road. In the past this would have been possible too, but the user would have to know the relevant cadastral numbers, parcel numbers, sewer identifiers and so on. This may be specific information that the user does not usually know, but he does know the area that will be affected by the work carried out on the road. Not only is the geographic information system an additional source of information, it also provides an additional method of accessing and presenting information. Provided that the operational use of the GIS is easy enough for occasional users, the benefits are not only those of being able to generate combined geographic and administative information, but also of having an additional data access method.

7.5 The implementation of a fixed asset management system

Before a fixed asset management system is implemented, it is necessary to take account of the social impact, technical issues and the problems and benefits which may arise. Any change always has its reactions. This is true when implementing a large computer system. People will react differently. Some will welcome a new means of doing their job, and recognise the benefits. Some will be indifferent and will reserve judgement until they have seen proof of better or faster information. Other potential users will be against change for various reasons. Technically, we are dealing with sophisticated computers requiring modern infrastructure.

Installation will sometimes mean building refurbishment. The new technology will also require training and assistance both in working with the systems as well as in setting up the user environment.

The social impact of new and modern technical products is sometimes underestimated. As already mentioned, fixed asset management will very often involve dealing with users who have not used computers before. If not managed properly this can result in only limited acceptance or, in some cases, even deliberate misuse of the system. There are a number of requirements in order to prevent a negative response. Firstly, management commitment is necessary. Secondly, potential users should be involved in the product at an early stage. In this way, the potential user will get to see the benefits of a production system and its acceptance will be less difficult.

The third requirement is user training. Initially it will be possible to train relatively large groups, but as soon as actual product training is required, small groups are essential in order to give each user enough time to familiarise himself or herself with the system. Training should not be the product of technical training alone. Some education about the aims and objectives and the philosophy behind the system is needed as well. A fourth requirement is the need to assure employees of their future job security and responsibility. It is also necessary to encourage pride in ownership of a GIS and careful negotiation within departments about which data is incorporated in the central data set. Users often tend to think that their own data sets are the most accurate and complete and sometimes choices have to be made. Caution is needed since disqualified data can be seen by the involved users as a judgement on their work. Finally, employees will be most motivated if their work reflects directly on the work of the department. Work that, in their point of view, only benefits other departments will be more difficult to have accepted. In order to fully deal with these social effects, a steering committee is required to guide the whole project.

The easy part of implementing a GIS is the actual installation, even if some building reconstruction is required. Once the system is installed it has to be customised to meet the users wishes. Good authorisation has to be designed to deal with both spatial and administative information, and users will have to be further trained to exploit the systems as effectively as possible. Since all users will be accessing the same central data, it has to be an absolute priority to make sure that the definition of this central data base is identical for each user. Great care has to be executed in building the database in order to guarantee a complete and accurate database. Nothing is worse than an data set with errors, because users will lose faith in the system. Provided that a good data set has been created, time should be spent, with each additional application, to tune the data even further. All users have to clearly understand the data definition. For instance, the definition of a persons name has to be absolutely clear. This may relate to how the name is spelt, whether it is in upper or lower case, and whether first initials, title and abbreviations are used. New users should clearly understand the guidelines and definitions so that optimum use of the database is guaranteed.

One of the main problems is that individual users will tend to be more restricted with a total application package that is suitable to all users. For example, a certain group of users may have always used a certain number code for streets, but in order to make the system work for all users, this code has to be changed. Typically this might occur in a surveyor's office. In order to make use of the graphics (i.e. the continuous map), relations will have to be stored with each graphic element. Therefore each house will have a unique building number and each street will have a unique street code. This will require a different working procedure in the surveyor's office, sometimes making the work more difficult but resulting in benefits for other departments. In order to implement a system, the benefits must inevitably outweigh the problems.

The benefits of implementing a GIS are multifold. Firstly, all users will be able to use the same central data set, thus eliminating multiple updates. The central data set will be more accurate and far easier to maintain. All users will have access to the map, no longer as a paper sheet, but as a computer presentation tailored to their requirements. This will eliminate paper copies, outdated versions and labour intensive work, and enable the user to present more readable information by using graphic presentation. However it is the tying together of each separate application through a spatial component that really provides the new and previously impossible benefits.

Each department will draw benefits from the automation of its current tasks and activities. However using a GIS for fixed asset management will provide additional benefits. Because all users access the same central data set containing both adminstative and geographical data, less time is required to maintain this information, and better guarantees can be given to secure accurate and up to date information. Due to the labour intensive activities required to produce and maintain the administrative and geographical information, the saving is far more substantial. A good GIS will use the database to display the spatial information. This will result in information that can be used. Typically when using standard maps in the past, differences in presentation, scale and accuracy prevented users from benefiting from information available in other departments.

Information has not been properly related in the past. Road pavement construction followed by sewerage and water supply maintenance can mean road works appearing at the same place two or three times in a short period of time. There are many other examples which illustrate the problems that occur when information is available to only one department and not to others. A GIS will make that information available in a useable form for all departments. This will create a new form of access to information. Combined with spatial analyses, it will enable new means of managing fixed assets. A GIS makes it possible to identify all road segments that are in need of repair combined with all sewer pipes that also require attention. Analyses like these would provide a better and more efficient service to the public. But this is only the beginning. Elements affect each other. Trees will grow better or worse depending on the road surface. Discharge from a leaking sewer may influence the health of a tree, but when a sewer is repaired this may also influence the surrounding vegetation. This means that for the green department, it is of great importance to know what other activities are happening.

Alternatively, the fire department can carry out risk analyses since information regarding the location of people and buildings related to the object to be researched are available. Spatial analysis will form the major benefit for all departments, allowing the government to be more efficient and to manage their assets more effectively. Spatial analysis is the key to proper management information, because data will be linked to create new information about particular locations, and maps can be drawn for convenient display.

We have assumed so far that all information is stored within the fixed asset management system. However, the reality is different. Computer systems are already employed to perform the other tasks. In order to use this data the fixed asset management system has to be able to link to these systems, so this information can be made available too. This will provide benefits for the fixed asset management users as well as for the other users. New information regarding occupants will be available so that users can match their activities with the needs and requirements of the population. Knowledge of the age and the number of children can help in selecting the best location for schools. Areas that have a high number of elderly people could be selected and policies to improve demographic balance can be formulated. The benefits would be felt by the departments primarily using the GIS as well as the department using other systems. A further step towards optmizing the use of the GIS will be to use data collected from external organisations, such as the Central Bureau for Statistics. However, in order to fully integrate a GIS into this environment, the system should have facilities to communicate with the other systems. An open systems platform is recommended since it will be able to adapt to almost any existing computer environment.

7.6 Conclusion

The implementation of a fixed asset management system requires a lot of work. The organisation concerned needs to be properly managed in order to make implementation of the system a success. It will require all the participants to focus on the goals, and sometimes will give the impression of being more beneficial to the overall organisation than to the separate departments. However, if implemented correctly, the system will provide benefits to each department as well as to the organisation as a whole. Once the data is available, the potential use is enormous. Physical planners are able to obtain all the information required to plan major changes to their cities. With more and more traffic and activities, this information will become increasingly crucial. The gathering, analysis, manipulation and presentation of data are bringing new dimensions to local government planning.

Peter H.A. de Gouw
Unisys The Netherlands N.V.
Hoogoorddreef 11
1101 BA Amsterdam
The Netherlands

8 GEOGRAPHIC INFORMATION SYSTEM DEVELOPMENT IN TACOMA

Stearns J. Wood

8.1 Introduction

The City of Tacoma, located thirty miles south of Seattle, Washington, is a dynamic urban area. The economy expands and shifts, people move and land use patterns and administrative boundaries change. Currently, it has a population of 160,000 individuals living in 72,000 households in an area covering 49 square miles. City government has the job of managing resources in this dynamic environment. It provides services, collects revenues, and promotes sound development practices and other activities beneficial to the health, safety, and well-being of residents. The management of a dynamic urban environment requires and generates large amounts of data. Changes in the City place daily information processing burdens on each of the City's departments. Each department needs up-to-date information on a daily basis to meet its responsibilities to the public, and departments must exchange data with other departments so all can do their jobs effectively. Much of this information is geographically referenced and can be automated. This type of data can be incorporated as locational data in a Geographic Information System (GIS).

In order to make effective use of the large volumes of data, automated systems have been adopted which are providing a key part of the answer. This paper presents the Tacoma Planning Department's approach in using Environmental Systems Research Institute's (ESRI) ARC/INFO software to develop a citywide Geographic Base System (GBS) to provide mapping and reporting, and analytical capabilities for City officials, managers, planners, engineers, and staff in all City departments that use geographic information to make decisions and to serve City residents.

8.2 The background: from GBS to GIS

The City of Tacoma has processed and manipulated geographic information for many years. These efforts began as early as the mid-1960s when the City automated its land use data for each City block. In the mid-1970s, Tacoma's GBS started out as a set of applications residing on an IBM mainframe to process and utilize an automated parcel map file. It has evolved in the 1980s into a geoprocessing system using ESRI's ARC/INFO software residing on a Prime 9955/II super mini-frame. ARC/INFO provides Tacoma with computerized capabilities for a parcel ownership inventory, land use planning, environmental mapping, property mapping, incident mapping, routing, spatial analysis, facility and resource management. The Tacoma GBS experience is undoubtedly similar to that of many other governmental jurisdictions, if not also to many businesses that use geographical data and maps. Tacoma is where it is today mainly because of the increasing need for urban information to make informed decisions; but more recently computer technology has been used to address the increasing work load of departments with limited and decreasing resources.

Parts of the text and graphics of my work have been printed or shown earlier in City of Tacoma reports or documents as part of the City's work program activities, and have been used in speeches and presentations, and consequently could be in part in the public domain of the United States of America.

H. J. Scholten and J. C. H. Stillwell (eds.),
Geographical Information Systems for Urban and Regional Planning, 77–92.
© 1990 *Kluwer Academic Publishers. Printed in the Netherlands.*

In 1973, the City of Tacoma and Pierce County joined together to research the development of a geographic base file. During the past 15 years, there have been a number of times of transition when GBS configuration, definition, and extent of development have been addressed. From 1974 to 1980, Tacoma developed a geographic base file containing ownership parcels and streets, as well as application programs (from GRAFPAC II, Wyandotte County, Kansas) to map, summarize and maintain data. In 1980, a Geoprocessing Systems Review Committee was formed to review, coordinate and oversee the City's geoprocessing activities, to ensure maintenance of the geographic base files, and to make recommendations on the effective provision of geoprocessing services. During 1982-84, Tacoma purchased PIOS, then ARC/INFO to update and maintain the geographic base file and to increase the production of maps and data. The Public Works Department leased a Computer-Aided Drafting (CAD) system (TERAK) in 1984 for map drafting, whilst the Planning Department purchased a Mackintosh computer to undertake urban design graphics. In 1986, members of the Public Utilities, the General Government and IBM Corporation formed a team to assess the need for a GIS for their respective jurisdictional areas. The previous systems had had impressive results and more departments had become involved, purchasing AutoCAD systems to improve map drafting even further. Thus, in the last few years, the City of Tacoma has taken further steps forward with GIS with the intention of improving the management of its data.

8.3   The geographic base system

The establishment of the GBS has provided Tacoma with the capability for handling geographic data entry, storage, analysis, and mapping more rapidly, inexpensively, and accurately than the traditional non-automated methods. While effective record-keeping, analysis, and management of information are important benefits of the GBS in daily operations, the greatest benefits of the GBS are those resulting from its application in support of decision-making. The GBS quickly and accurately provides decision-makers with information. The GBS readily produces maps and supporting tables of information that used to be difficult or time-consuming to produce. It makes a multitude of relationships in information from varied sources immediately accessible to decision-makers, who can then spend their time debating policy rather than assembling facts hidden in volumes of written information.

GBS development objectives

The primary development goal of the GBS has been to implement a system (hardware, software, data processing staff, users, operational procedures, geographic data) which provides a computerized capability to store and account for all land-related information in the City, and to make it available to all City users. To date, this has been accomplished through use of the ARC/INFO System, geographic base files, the parcel level ownership file, and user and general application files.

The preliminary objectives of system development have been to provide adequate and timely data and maps for City operations, to provide data and maps on short demand to customers, citizens, departments and agencies, to encourage employees to use the computer in accomplishing their work tasks, and to encourage the involvement of all departments to automate their geographic data and use the geographic base files and ARC/INFO to produce maps and generate data. In this effort, Tacoma has been somewhat successful.

GBS organization functional interrelationship

The City's departments and agencies relate to and perform the primary functions for which the City has responsibility in carrying out its duties in serving the public. All departments and agencies normally have need for geoprocessing services especially when working with locational data. Most have frequent need for GIS support where others have only occasional need. Figure 8.1 shows the basic functions of government in providing services to the community and how the GIS organization is meant to interrelate these functional activities for data and map exchange.

Figure 8.1: Planning functions relating with the GIS

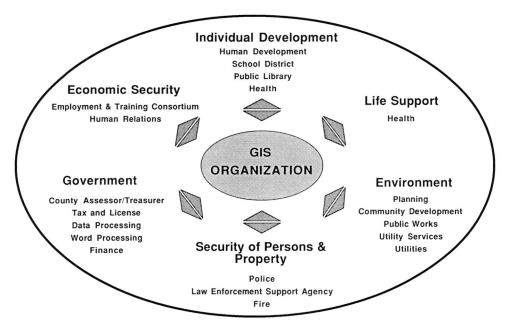

GBS schematic framework

Tacoma is using ARC/INFO GIS software to develop its citywide GBS to provide map and data products to all departments including the support of the City's planning process. A key element of the GBS system software is its geographically-based data management capability integrated within the mapping function to manipulate and display maps or geographic features (such as points, lines, and polygons). The database manager organizes the attribute data tied to the various map features. The data in the geoprocessing system is all related to geographic features. These data are either specific x and y coordinates indicating the location of those features or attributes associated with the features. The geographic features may be points, such as the specific location of a traffic accident or fire hydrant; lines, such as street or water line segments; or polygons, such as parcels, blocks, and planning, police or fire districts. Figure 8.2 shows the base map and the other associated data sets which are related through GIS.

The geoprocessing system, through data management, includes the capabilities to store all geographically related data in a manner in which they can be related to their geographic location and can be retrieved, processed, reported and plotted in the context of geographic location. The system also enables the user to relate data associated with one or more types of geographic features with data associated with other features to perform simple or complex analyses.

A distinction needs to be drawn between GIS functions and applications. A system function is a particular operation of the GIS software. One or a combination of system functions may be used to perform an application. Often, commonly used applications are performed by stringing system functions together through the use of a macro program. Some examples of full GIS system functions include: graphic and attribute data entry; overlay of map layers; data editing and deletion; map scale change; graphic display; map projection transformations; plot generation; merging of database components; graphic and attribute database query; edge matching; geographic proximity search by attribute or graphic element; report generation; digital terrain modelling; coordinate geometry routines and engineering calculations; dimensioning.

Figure 8.2: The geographic database model

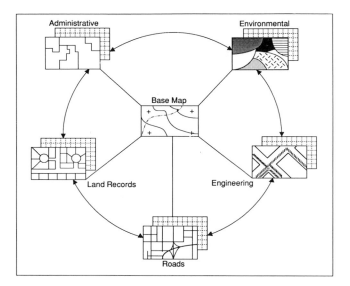

GBS database hierarchy

The levels of data within the GBS database are as follows:

FIRST
- control level definition
- $\pm$ .01/.001 foot resolution
- base mapping, surveying, Global Positioning System (GPS)

SECOND
- facility level definition (base map and engineering)
- $\pm$ 6 inch resolution
- base mapping, AM/FM - CAD, surveying (lot and block lines, planimetric base map, orthophoto map; electric, water facility distribution)

THIRD
- parcel level definition (land records and roads)
- $\pm$ 3 foot resolution
- geoprocessing - spatial analysis (property addresses, alias addresses - misspelt + common name; property ownership parcels; street and alley right-of-way parcels; street centerlines, street intersections, street networks; community/public facilities; community incidents)

FOURTH
- area level definition (administrative and environmental)
- $\pm$ 20 foot resolution
- geocoding - geoprocessing - spatial analysis (city blocks, census boundaries; administrative district and jurisdictional boundaries; natural systems)

GBS base files

The geographic base files of the GBS include:

(i)   The map definition file (Librarian System) which contains the definition of which quarter section maps make up the extent of those maps needed to identify administrative, jurisdictional or district areas.
(ii)   The coordinate points file which contains the x,y coordinates for every point in the City that has been identified as well as control level definition coordinates.
(iii) The parcel map file which contains information on every parcel in the City along with the points that define the parcel.
(iv)   The area boundary file which contains the point numbers for overlay polygons such as fire response network and zones, commercial and industrial business districts, police districts, development intensity areas, planning areas, zoning and shoreline districts, census tracts and census block groups, council districts, neighbourhoods, Indian reservation boundary, school boundaries, natural systems areas, and so on.
(v)   The alias address file which contains special addressing for identifying and locating properties which have several or unique addresses assigned, and recording frequently misspelt street names, and for recording properties which are known by a familiar name (e.g. Point Defiance). Personnel identify streets with more than one name, streets in different areas of the City with the same name, streets with different name spellings, and streets commonly misspelled. This is accomplished by processing operational data files of the user application departments through the ADMATCH program for address editing. The street name errors are extracted and placed in a central file.
(vi)   The street centerline file which contains the coordinate definition of street networks through the City, the format of the street name, and the associated address ranges.
(vii) The street intersection file which contains the x,y coordinates of each intersection in the City along with the formated names of the intersecting streets.

Figure 8.3 shows the map layering concept of a GIS map database presenting the hierarchical relationship from the base high resolution control level up through the lower resolution area levels. Figure 8.4 outlines the schematic framework of the relationship between the core reference files of Tacoma's GIS and the various other system and application software and database segments of the GIS, especially user application programs and database files, and system software and support systems programs.

GBS database maintenance

In order to provide data to maintain the GBS files, data from many departmental operations is utilized. Following are the governmental operations which are currently accessed for GBS purposes:

(i)   Assessor - tax rolls, property evaluations, parcel ownerships;
(ii)   Public works - surveys, subdivisions, short plats, building permits
(iii) Storm water utility - service changes
(iv)   Public utilities - customer information system
(v)   Planning - land use surveys, permits
(vi)   Police - calls for service
(vii) Tax and license - business openings and closings, licenses
(viii) Fire - calls for service and inspection

Figure 8.3: Map database layering concept

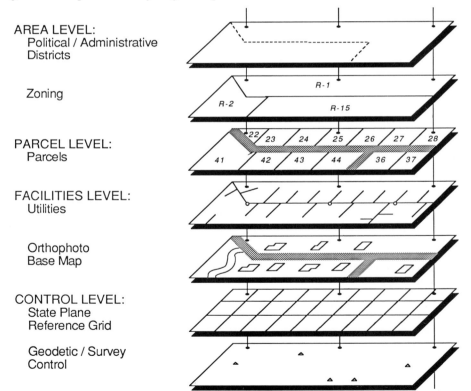

Figure 8.4: Schematic framework of the core GIS

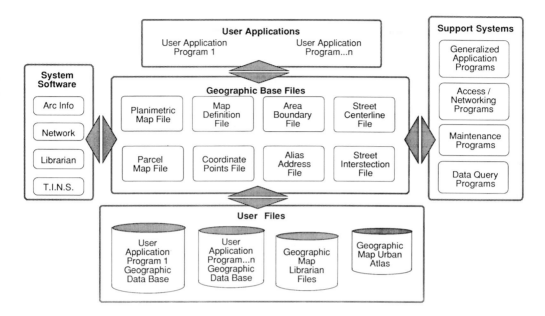

GBS data files

Computer data files when interrelated with automated map processing can benefit many governmental and private organizations. Some possible data files of an urban area for use in geoprocessing include:

(i)   Land: topography, rivers and flood plains, geology and soil types, land and water resources, property ownership (tax assessment);

(ii)   Use of land: buildings (inspection and licenses), trafficways, water areas, open space, urban and rural use categories for land planning such as residential, commercial, industrial, and so on;

(iii)   People at locations: population, housing, and census demographics; school children, business firms and employees, trips related to origins and destinations; election statistics;

(iv)   Events at locations: locations of fires, crimes, accidents, welfare, health, and hospital cases; location of customers, levels of market usage (market research); sources and levels of air and water pollutants;

(v)   Facilities at locations: location of street lights, fire hydrants, sanitary and storm sewer lines, water lines, electrical distribution, cable television, telephone and natural gas lines;

(vi)   Boundaries of special areas: geo-political, jurisdictional, service and/or physical boundaries; City limits and planning areas, land development intensity and land use zoning areas, shorelines, police reporting districts, fire response areas, watersheds, neighbourhoods.

GBS data exchange

Data and computer maps are exchanged with other jurisdictions and agencies which need to process information about the Tacoma area. Examples are Environmental Protection Agency, United States Geological Survey, Puget Sound Council of Governments, Pierce County, Tacoma School District and the State of Washington. Figure 8.5 shows the detailed relational definition of the computer hardware and software with the database files, the user application programs, and the general multiple-user application program segments of the GIS.

Figure 8.5: Schematic framework of the GBS

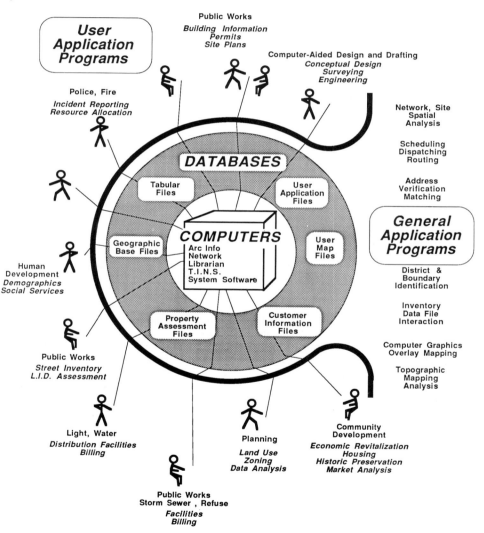

GBS user applications

A number of City departments (Planning, Public Works, Fire, Police, Community Development, Human Development, Public Utilities) have been using the GBS, as well as other agencies such as Office of Intergovernmental Affairs, Tax and License, Health, City Manager, and the Pierce County Office of Assessor/Treasurer. The computerized programs undertaken by these departments and agencies to satisfy their operational, analytical and geographic needs represent the GBS user applications. The Tacoma Planning Department, for example, utilises systems for land use information mapping, zoning information mapping, census information mapping, parcel information inquiry, on-line geographic base data management, and office automation.

Most of the agencies and departments having GBS user applications have activities which are similar and can have common programs developed which can be used by each. General applications are developed and placed in the GBS so that any agency or department with a similar activity can utilize the centralized computer program. Examples of such general applications of the system include: land/facility inventorying, district and boundary identification, networking, geocoding, address verification and matching, site/network/spatial analysis, incident reporting, computer graphics/overlay mapping, topographic mapping/analysis, scheduling/dispatching/routing, on-line information retrieval/user data file interaction, and data base management.

An example of area study mapping of the Planning Department's land use planning application using the GIS parcel map file to produce data and maps for the City Waterway Design Plan is illustrated in Plate 8.1. Another example (Plate 8.2) shows line segment mapping of the Planning Department's land use planning application using the GIS street centerline and parcel map files to produce data and maps again for the City Waterway Design Plan. Plate 8.3 shows an example of the use of one computer system (Tacoma's ARC/INFO GIS) by more than one local jurisdiction for GIS application use (to produce a generalized land use map of Pierce County, Washington). Plate 8.4 shows an example of the joint use of one computer system (Tacoma's ARC/INFO GIS) by two local jurisdictions for GIS application use; in the Spring of 1990 the City of Tacoma and Pierce County will obtain exact positioning of monuments through consultant use of satellite systems from the Global Positioning System.

GBS organization and policy

In 1980, the Geoprocessing Systems Review Committee, consisting of persons from six user departments plus data processing, was established to oversee the GBS. This was expanded to eleven departments plus data processing in 1986. Geoprocessing service policies were set up defining geoprocessing and GBS, user interrelationships, geoprocessing system organization and responsibilities, use and maintenance cost-sharing formulas, and an intergovernmental information sharing agreement. In 1986, a GIS Task Force was set up to investigate, identify and recommend the development of a true GIS and a Management Group was established to oversee this effort.

GBS development philosophy

The philosophy and approach Tacoma has taken to develop the GBS has been to promote the need for geoprocessing, and to collectively start modular systems development toward implementing a total integrated system. When user applications are started in an agency, they are developed in an operational environment satisfying the day-to-day needs of the departments. They are designed to respond to the daily and ad hoc demands of the departments for the information needed to accomplish routine operations. This approach has helped to introduce GBS into the governmental framework, as part of normal data processing developmental and production costs. This has helped to show early GBS results, provide the sources for maintenance of the GBS, and provide substantial footing for expanding into the analytical and computer graphics environments. Figure 8.6 shows the various aspects and relationships of the administration of Tacoma's GIS in that it serves two separate organizations (General Government and the Public Utilities) and their many departments and divisions.

Figure 8.6: GIS administration in Tacoma

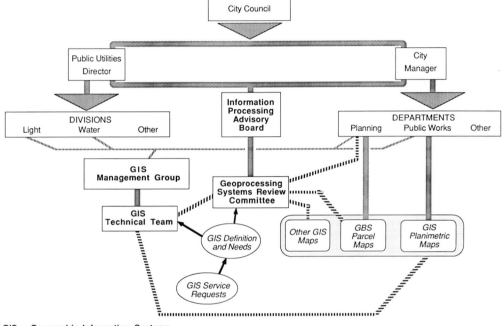

GIS = Geographic Information Systems
GBS = Geographic Base System (Geobase)

During the development of Tacoma's GBS, the following development philosophy and approach has been adopted:

(i)   Obtain administrative support
(ii)   Obtain inter-departmental and inter-agencies support
            - lead departments and agencies assume responsibility
            - share update and maintenance
            - undertake coordination and integration
            - form GBS committee to guide development
(iii)   Obtain GBS financial support; cost justify to management

(iv) Maintain technical staff expertise and resources
        - maintain GBS technological sophistication
        - train employees in GBS technology
(v)   Promote development and use of GBS
        - operational environment, daily task performance
        - promote user department experts
        - encourage access and availability of GBS
        - make GBS customer oriented
        - promote fast pace organization with innovation
        - adapt to and build in change
(vi) Encourage favourable GBS operational climate
        - project best possible utilization of GBS
(vii) Seek interagency (public and private) support
        - all local governments
        - federal, state and regional assistance
        - seek technical assistance
(viii) Improve existing mechanisms and processes
        - operational to analytical and graphics
        - old technology to new and improved
        - advanced data processing capabilities
        - innovative research and development
        - single to multi-purpose and integrated
(ix) Begin GBS development, involvement and accomplishment
        - undertake modular development, have pilot projects
        - obtain visual results
        - stress quality and accuracy

GBS benefits of geoprocessing

During development of geoprocessing in the City, a number of benefits have been realized by users of the GBS. The type of benefits that have been and can be derived through use of GBS and geoprocessing are as follows:

(i)   Rapid and coherent access to data: obtaining data, maps and reports quickly and in an understandable form as compared to the time it takes to do the work manually.
(ii) Integration and interfacing of data: obtaining the data needed no matter in which department the file resides and getting the desired data in the format needed to produce maps and/or reports.
(iii) Decreased duplication of effort: reducing duplicate effort by departments sharing automated data files and centralizing computer base maps or common use maps into a single location to be used by each.
(iv) Improved reporting capabilities: getting the latest accurate information in the format needed and using GIS to explore previously unavailable areas of spatial and site analysis.
(v)   Improved accuracy and resolution: correlating data sources and providing improved means for data editing and correcting, resulting in greater freedom to identify error or mistakes and reducing processes to simpler form.
(vi) Dynamic updating of data: updating the GBS files automatically on a daily basis in an operational environment so that the latest data is always available.
(vii) Decreased manual effort: saving personnel time by having the computer do the work of compiling data and producing reports or maps automatically. By eliminating basic map and report functions previously performed by hand can, through automated mapping and

GIS, improve speed and accuracy to free personnel to do other planning tasks and relieve the tedium of mechanical tasks.

(viii) Improved productivity: realizing all the above benefits with employees accessing the geoprocessing system as a working tool to perform more efficiently, providing a better and more reliable product in a shorter time period.

8.4   GIS concept

The recent Tacoma GIS studies have recommended that the City's GBS be expanded to make it operationally functional for all users in the City's General Government departments and its Department of Public Utilities. As a direct result of Tacoma's early participation in GIS, the application of the technology in the City has evolved along a parallel track with the industry. Departments with a primary CAD focus have acquired AutoCAD, and departments with a focus on geoprocessing and map analysis have developed using GBS and ARC/INFO. ARC/INFO and AutoCAD, are both state-of-the-art solutions for their primary disciplines.

The two major barriers between the City's current GBS system and the proposed citywide GIS are the lack of Automated Mapping and Facilities Managment (AM/FM) functionality, and the lack of an adequate Data Base Management System (DBMS) including system integration or connectivity. As the City's systems have evolved, neither ARC/INFO nor AutoCAD have gained universal popularity; both are currently limited in terms of AM/FM applications and an integrated database management capability. This lack of the AM/FM component, as well as the lack of integration between the City's systems, are the key considerations for future GIS configuration design.

GIS functionality

Although there is not a universally accepted definition of GIS, the functionality and activities traditionally associated with GIS are well defined. As viewed by a majority of current users, a true GIS is a system that incorporates and integrates the functional sub-system aspects of CAD, AM/FM, graphic spatial analysis of geocoded data (geoprocessing), and connectivity to large, main-frame based DBMS. The hallmark of a true GIS, however, is its geographic intelligence - namely its topology. Topology is a description of the specific relationships used to represent the connectivity, contiguity, or proximity of the geographic features included in a map. By storing information about the location of a feature relative to other features, topology provides the basis for many kinds of geographic analyses.

In 1975, when Tacoma first went into GIS, the disciplines of CAD, AM/FM, DBMS, and geoprocessing were all separate products. There were no products that successfully integrated functionality from all four disciplines. Because of this, the City implemented just one of these areas, geoprocessing, as the GBS. For the Planning Department, and agencies such as Police, Fire, Community Development, etc., ARC/INFO has been very successful. For Public Works, Light, Water, and Sewer Divisions, AutoCAD now serves as the primary computer-aided drafting tool. To this use, AutoCAD has been very successful.

There are valid reasons why one system or the other has not gained universal popularity and become the citywide GIS that Tacoma needs. These involve limitations in functionality, especially in the realm of AM/FM applications. The main benefit of AM/FM to all users is a refined base map covering the entire City which includes planimetric and facility-related information.

Existing GIS are rooted in one of the three earlier disciplines and show unbalanced strengths and weaknesses with respect to one or another of these disciplines. Some, however, are evolving toward the comprehensive functionality that characterizes true GIS. Tacoma needs to identify and select a GIS concept which successfully integrates these functional disciplines. This it is currently doing. Use of these systems within the City can be successful whenever adequate development support, user training, and actual need exists. There is no doubt that the use of the above systems can and should be expanded to encompass a wider range of applications.

Figure 8.7 shows the system functionality of a GIS that incorporates and integrates the sub-system aspects of CAD, AM/FM, graphic spatial analysis of geocoded data, and database management system/ interconnectivity. GIS network connections between Tacoma's current IBM computers, the Prime computer which holds the ARC/INFO system for geoprocessing, and the AutoCAD computer system for CAD, will therefore be extended to include the proposed AM/FM system. AM/FM is a dedicated computer system that stores the location and description of all the equipment and facilities of a utility or public works department. Included are all the map backgrounds, the structures that hold the equipment and facilities and the annotation for them. In addition to pictorial data, information describing the equipment and facilities is stored. The key to AM/FM is the automated integration of graphics with data for all equipment and facilities.

Figure 8.7: The integrating concept of GIS

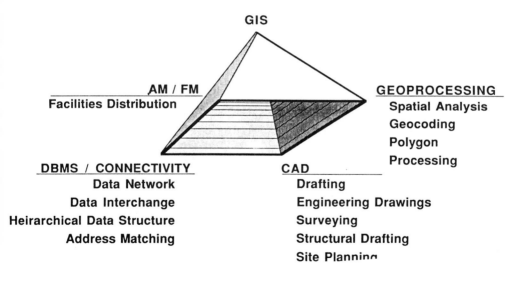

AM/FM systems provide the engineer with a means to create and maintain a database model of a utility's or public work's distribution system or facilities network. AM/FM systems provide the agency with the means to automatically draw, maintain, and retrieve information from equipment and facilities location maps. This helps to relieve the awesome and time-consuming management task caused by manual operations and to reduce map and data inaccuracies, redundancies, conflicts and data voids as a result of changes in standards, technologies and growth patterns of utilities and public works departments.

Extension of the citywide GIS to encompass AM/FM provides the City with additional functions such as the mapping of public facilities over a common base map and of a continuous surface covering the entire service area, and the inclusion of extensive facilities attribute data in the database. It also permits facility modelling on a large scale, and more expedient utilities and public works facilities data sharing with other agencies.

The DBMS is a critical part of the GIS. A database must contain coordinate location data, non-coordinate location data (e.g. address), topological data describing relationships between locational features and attribute data related to the geographic locations. A database must be structured in such a way that the relationships among locations and between locations and attributes may be easily recognized. It is necessary to incorporate and integrate all geographic data into an overall City, user defined DBMS. Developing a DBMS addresses the objectives of exchanging data between all City departments easy and efficiently. The database is the repository for the data of the geoprocessing system. This data is obtained from a variety of sources within and outside the City organizational structure. As a common repository of data for all users, the database significantly enhances the exchange of data between users and therefore directly improves the efficiency of City operations which are geographically based. With respect to database design, data integration means more than just daily sharing. The ability to transfer files across functional platforms is not sufficient to define an integrated GIS. Data must be integrated into a common database, and this database must be accessible to all users. True multi-user access is required. Users must be able to simultaneously view the same data in the same geographical area of the database.

GIS system development

The recommendations of the 1989 GIS consultant study are that the City should:

(i)   implement a phase I centralized system configuration that consists of AM/FM and DBMS/connectivity which can be linked to existing AutoCAD and ARC/INFO systems in a phase II distributed configuration;

(ii)   create a digital quarter section planimetric base map that contains all data from the City's official map: curbs and road edges, lots and blocks, driveways, building footprints, hydrology, sidewalks, topography with five foot contours, etc.;

(iii) undertake an aerial photography flyover to obtain needed base mapping information and accuracy in horizontal and vertical survey control and to prepare orthophoto maps;

(iv)   develop a Digital Terrain Model (DTM) to provide the ability to develop and manipulate contours;

(v)   develop a new GIS Central Coordinating Division with a GIS Manager and supporting GIS staff with responsibilities to develop and maintain the City's common base map and to implement the new GIS processing capability; and

(vi)   create a new GIS Management and Policy Group and a GIS Technical User Group to oversee the citywide GIS.

GIS base map description

The GIS study provided an inventory of 125 map sets in use within the City. Of the map sets in use, most of these are maintained by Light, Water, and Public Works and only a few are automated. The identification of these map sets provides the assessment that there is a high level of redundancy in mapping activities. With some exceptions, most of these are based on the City planimetric quarter section maps and contain the majority of the facilities maintained by the departments. The number of special map products produced and maintained by each department is significant. The reason for the high number of these maps is due largely to the manual mapping processes.

The City's departments have a wide variety of geographic information that should be integrated into a common database accessible to all users. For this, Tacoma needs an improved base map with improved accuracy before the City's maps can be converted to a single automated citywide map. Planimetric quarter section maps are the City's 'official map', manually produced and maintained by the Public Works, Engineering Division. These are considered the most spatially accurate maps in the City. The maps contain the following: public land survey monuments and lines; Washington State plane coordinates for monuments; street centerlines; centerline bearing and distance, and curve data; street rights-of-way; right-of-way dimensions; street names; alleys; deeds, easements, judgements, resolutions; water features; railroads; plat names and numbers; plat/lot boundaries; lots and blocks and their dimensions.

These maps are oriented and formatted to the quarter sections at 1" = 100' scale. The 222 quarter section maps cover Tacoma and portray the original legal divisions of land and do not necessarily reflect ownership. They are manually updated, and routed to 13 separate departments and agencies within the City. The maps are the City's primary 'base maps' and are considered the primary source for base map layers for the citywide GIS database. The improved base map will serve as the vehicle by which all departments can store, share, and compare data.

In order to implement multi-user access to common data, a common cadastral platform will be developed as the automated base map, and its digital conversion and maintenance will be given high priority. The potential use of GIS technology using such a base map within an environment such as Tacoma is nearly endless.

GIS application development

An application uses the database and system functions to retrieve, portray, and manipulate data in a particular manner needed to support the information requirements of the system user. It uses numerous graphic and attribute data elements and several system functions to complete a process. Sophisticated applications may involve complex mathematical computations, combinations of data, and portrayal of the results in both tabular and graphic formats. Potential GIS user applications identified for different departments by the study are numerous. Systems for the Planning Department alone deal with systems for land use permit information, geographic information referencing, urban design graphics, urban data collection, analysis and publication, capital improvement programming and asset management, and management information referencing.

8.5 Summary

Even though Tacoma has a GIS, it is constantly evolving and being improved on a daily basis. No one can purchase a true GIS and have a fully operational system immediately. GIS hardware and software can be purchased, but to have an operational GIS database it takes years of development. After 15 years, Tacoma's GIS efforts are still under development.

During the 1990s, the City's efforts will be to continue what was being done in the 1980s. Work will expand these activities into user-friendly processes where City staff, without technical computer knowledge, can 'pull down' combinations of data for any area of the City and provide map and data products quickly. Tacoma's efforts will also be directed at addressing the management of GIS as a single organizational entity, and expanding GIS as a transparent user tool into many more City operations and activities. This will include not only expanding geoprocessing but also interfacing GBS with CAD, developing AM/FM, and further defining and implementing the City's DBMS, all to provide a true geographic information system in a distributed environment.

Stearns Wood
Tacoma Planning Department
740 St Helens Avenue
Tacoma
Washington 98402
USA

PART IV

DECISION SUPPORT SYSTEMS FOR LAND USE PLANNING

9   REGIONAL PLANNING FOR NEW HOUSING IN RANDSTAD HOLLAND

Stan C.M. Geertman and Fred J. Toppen

9.1   Introduction

In 1988, the Dutch government published its Fourth Memorandum on Physical Planning (Ministry of Housing, Physical Planning and Environment 1988). This report sketches a broad perspective for urban and rural development in the Netherlands up to the year 2015. The Ministry of Housing, Physical Planning and the Environment is currently working out the details. In doing so, they wish to use the Geographic Information System (GIS) ARC/INFO to translate general policy statements into concrete location decisions. The GIS research group at the University of Utrecht was asked to develop and execute a method to support location decisions with the help of geographic information systems.

The central theme of this paper is to explain the process of developing a prototype GIS application to provide a system for supporting location decisions with respect to the implementation of urban and rural planning proposals. We will give special attention to the way in which such an application can fulfill the needs and expectations of its potential users. The application which will be described as a GIS case study is based on a set of criteria derived from the Fourth Memorandum. Criteria were selected to evaluate possible building sites and to support decisions concerning the location of additional residential areas in Randstad Holland (i.e. the extended urban conurbation including Amsterdam, Rotterdam, The Hague and Utrecht).

Firstly, the preferences, needs and expectations that planners in a policy environment hold with regard to GIS will be considered. A pilot study in which the prototype of a GIS application that provides a system for supporting location decisions will then be introduced. The method that was used to fill in part of the general planning perspective for urban and rural development will be illustrated. After a brief introduction to the study area, the way in which relevant goals and location criteria were extracted from the Fourth Memorandum will be described. Those goals and criteria are essential in evaluating possible building sites for housing (e.g. reinforcing the role of public transport or preserving valuable landscapes and nature areas). The construction of GIS map overlays from goals and location criteria will then be reported, emphasizing the combination and assessment of map overlays to conform with the urban strategy. The method that was developed will be discussed in detail, and some examples will be elaborated and illustrated with maps and tables. Finally, some of our experiences in developing and testing this GIS application will be presented. The conclusions are geared to the needs of planners working on the implementation of the global ideas put forth in the Fourth Memorandum.

9.2   The policy environment

Physical planning in the Netherlands is the task of national, regional and local authorities. They prepare, decide on and work out plans concerning the physical structure of the country. All kinds of people are involved in the process which comprises a broad range of

95

H. J. Scholten and J. C. H. Stillwell (eds.),
Geographical Information Systems for Urban and Regional Planning, 95–106.
© 1990 *Kluwer Academic Publishers. Printed in the Netherlands.*

activities. Needs for data and requests for information therefore differ accordingly. A general GIS like ARC/INFO can hardly be expected to satisfy such a wide array of demands. In this policy environment, four kinds of potential GIS users may be distinguished (Scholten and Padding 1988):

(i)   A group of information specialists who are professional users of geographic information systems, who regularly work with computer software and who are well versed in the methods and techniques of scientific research. A common GIS like ARC/INFO is, in their opinion, limited in its geographic analytical capacities. They desire more options for mathematical manipulation and model interfacing.

(ii)   A group consisting of people preparing and working out policy in a specific area of physical planning. They are usually not well acquainted with statistical or spatial methods and techniques. However, they need relevant spatial information and analytical methods. With the help of an information specialist and/or a user-friendly shell, they can obtain the desired information without needing to master the command language of the GIS themselves.

(iii)   A group of political decision-makers whose knowledge of automation and statistical methods is relatively limited, yet whose need for a decision support environment is increasing.

(iv)   A final group comprising individuals, other authorities, and interest groups who need to be kept up to date on physical planning policy and its ramifications for their sphere of interest.

The planners working in the policy environment we are concerned with here (ie. policy to implement the ideas put forth in the Fourth Memorandum) form part of the very large second group of potential GIS users. They cannot make independent use of a common GIS because of the complexity of the GIS command language. To remedy this situation, a user-friendly GIS application should be designed and implemented. With a user-friendly application, those who are not GIS specialists would be capable of manipulating at least the elementary functions of a geographic information system.

One of the tasks of this group of planners is to design alternative locations for about one million new dwellings in the Randstad area. To serve their needs, a GIS application should support the analysis of goal achievement and the process of design. Specifically, it should meet the following requirements (Geertman 1989):

(i)   it should be flexible in order to answer slightly different questions;
(ii)   it should be integral and efficient; i.e. capable of communicating with other applications and existing information systems;
(iii)   it should be interactive, which means that direct communication is possible between computer and user; questions and answers should be clearly formulated (user-friendly) and response time should be quick (need for work stations);
(iv)   it should be effective; i.e. capable of performing functions like queries on spatial elements and their attributes; it should execute spatial analysis functions like buffering and overlay; and it should have drawing capabilities for on-screen design of potential building sites. Actually, the drawing capabilities of a GIS are poor and a CAD system is more appropriate. However, the only way to make use of extensive databases and analytic GIS functions is with a geographic information system.

In general, only some of these demands can be satisfied by the commonly available geographic information systems. That is why the Ministry asked the GIS research group at the University of Utrecht to design and implement a method to support the work of people who are not information specialists in the preparation and execution of location decisions.

This method should also generate a new GIS application to be constructed around the existing ARC/INFO system. Our research at the GIS laboratory in Utrecht has emphasized the systematic development and evaluation of GIS applications, as well as the incorporation of more fundamental geographical methods like interaction and network analysis in an existing GIS (Harts and Ottens 1987, De Jong and Ritsema van Eck 1989). The outcome of this approach should serve the above-mentioned demands of this diverse group of potential GIS users. The process of designing and implementing such a method and the resultant GIS application will be discussed in the next section.

9.3 Case study

Introduction

The research project started in September 1988 and the first phase, a feasibility study, has been completed (Toppen and Geertman 1988); this section will present the results. In addition, it will briefly review the current phase of the project, currently half complete, in which we have built an operational GIS-based model for goals achievement analysis and design in the context of the selection and evaluation of possible building sites.

The main objective of the pilot project was the development and testing of a method of building a prototype GIS application to provide a system for supporting location decisions with respect to the implementation of the urban and rural planning proposals. The process of developing and testing the method of building the GIS-prototype is summarized in Figure 9.1. Some elements indicated in this scheme will be elaborated upon in the text that follows. Firstly, the study area will be described. Secondly, some details will be given of the supply and demand for housing in the period 1990-2015. Thirdly, some criteria derived from the Fourth Memorandum and pertinent to evaluating the location of building sites will be highlighted. The basic GIS application (Section I in Figure 9.1), which is focused on the evaluation of designated and alternative new building sites with respect to the criteria, will then be presented. Finally, some conclusions will be drawn which are important for the last phase of the research, the development of an user-friendly interface for this GIS application.

Study area

In the Fourth Memorandum on Physical Planning, the Dutch government stresses the importance of the Randstad. In order to compete with cities such as Frankfurt, Brussels, Munich, Paris and London, future economic growth has to be concentrated in this highly urbanized part of the Netherlands. Besides economic growth, population growth will also be concentrated mainly in the Randstad area (Ministerie van Volkshuisvesting en Ruimtelijke Ordening 1988). For this reason, a pilot study was commissioned to evaluate proposed sites and select new ones for residential development in this part of the Netherlands. From the Randstad area, which officially covers the provinces of Noord-Holland, Zuid-Holland, Utrecht and Flevoland, ten urban regions were selected for study (Figure 9.2).

Housing supply and demand

As indicated above, the Randstad will undergo significant population growth. In addition, changes in the size and composition of households will also occur, partly because of trends in lifestyle. These shifts will cause a demand for about two million dwellings in the Netherlands, of which one-and-a-half million will be in the Randstad region (corresponding to 20 percent of the existing housing stock). One million units are supposed to be built on new building sites; half a million will be produced by the renovation and

Figure 9.1: Outline of the pilot study

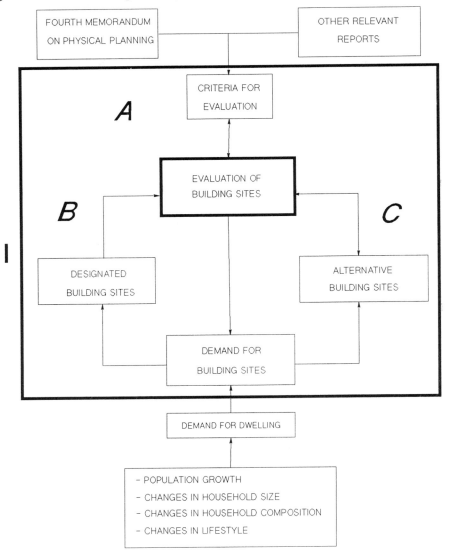

restructuring of existing building sites in areas already built up. This project is concerned only with the new building sites and more specifically, only those sites within the selected urban agglomerations in the Randstad area. Specified for this area, the total demand for the period 1990-2015 is almost 600,000 dwellings. To satisfy this level of demand, a number of building sites have already been proposed for development in the period 1990-2015. However, most of the supply will be available in the period 1990-2000. These designated building sites were previously identified in several physical plans of the regional and local governments. In Figure 9.2, the designated building sites are indicated for the study area. It is obvious that the designated building sites will not satisfy the total demand. The mismatch between supply and demand is about 250,000 units. It is also obvious that the shortfall is larger for the years 2000-2015. This is partly due to the fact that this period is beyond the scope of the current regional and local plans, in which those designated building sites are identified. In order to diminish the mismatch between demand and supply in both periods, the National Physical Planning Agency seeks to identify alternative new building sites. And new sites must be designed interactively, which is most effectively done with the help of an user-friendly GIS.

Figure 9.2: Built-up areas, urban agglomerations and designated building sites in the Randstad

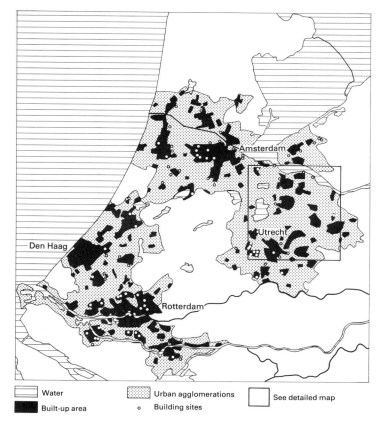

Criteria

The location of those new building sites has to be evaluated in terms of a number of criteria, which can be derived from the Fourth Memorandum and other relevant reports that deal with the spatial aspects of economic, housing, environmental, recreational, and rural issues. Since this was a pilot study, the criteria were not determined after an extensive selection procedure. They were selected for the insight they might provide into the feasibility of the method. It should be stressed that the results of the evaluation were not the main concern. Instead, we were most interested in the process of deriving criteria, together with testing the possibilities for combining them in different ways and giving the individual criteria different weights. In the subsequent phase of the research, together with the final programming of the application, the selection of criteria will be handled more precisely and in greater extent (Geertman and Toppen 1989).

In the Fourth Memorandum, the Dutch government formulated some global goals, two of which were especially of interest for our purposes. First, in order to reduce mobility, the government emphasizes the allocation of dwellings in the same urban agglomeration as the place of work. From this perspective, the government wants to promote the use of public transport within the daily living environment. Therefore, the location of residential building sites has to be related to the location of working areas, the built-up area, and service and recreational facilities. Second, the government's goal is not only to create favourable conditions for potential building sites, but also to put restrictions on such locations. Those restrictions may concern safeguarding existing nature areas and also protecting the Green Heart of the Randstad from large-scale construction activities. Therefore, building sites have to be evaluated in terms of the location and quality of agricultural areas, nature areas, large-scale landscapes, and greenbelts between built-up areas. Also, environmental aspects (pollution, traffic noise, etc.) have to be taken into account.

The translation of these global goals into criteria which can be expressed in more or less detailed map layers is the next step in the process of developing a GIS application. It is, of course, influenced by the presence, quality and accessibility of databases. Therefore, the selection of criteria was partly dependent on the availability of the corresponding databases. The following criteria were derived from the global goals:

(i)   from the goal of reducing mobility:
        -   new building sites must be adjacent to built-up areas;
        -   a building site must be within a certain distance of secondary schools (<3km, <5km); and
        -   the location must be within a certain distance of railway stations (<3km, <5km) (Figure 9.3)

(ii)   from the goal of safeguarding nature areas:
        -   no building sites are allowed within greenbelts (zones which are supposed to separate urban agglomerations) (Figure 9.4);
        -   no building activities are allowed in large-scale landscapes. (The scale of the landscape is translated into a measure of open space for each grid of 100 ha.); and
        -   no building sites are allowed in important nature areas. (The importance of a nature area is given as the number of hectares of nature for each grid of 25 ha.)

Figure 9.3: Areas within certain distances of railway stations and high schools

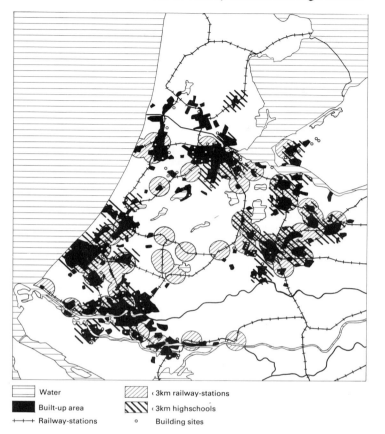

Water

Built-up area

+—+—+ Railway-stations

▨ ‹3km railway-stations

◪ ‹3km highschools

○ Building sites

The GIS application

In the application, designated building sites have to be evaluated against the background of the criteria selected. Also, additional building sites must be identified to fulfill the remaining demand for new sites and for evaluating these locations. Therefore, a user-friendly GIS application is desired to assist planners in presenting alternative new building sites by way of on-screen digitizing and in evaluating them. On-screen digitizing requires a topographical map and some criteria as a background. Alternative building sites can be digitized, and then other criteria will be used to evaluate the results.

This GIS application may be summarized as indicated in section I of Figure 9.1. The application consists of the following elements (the letters A, B and C referred to in Figure 9.1):

A (i)   select criteria;
  (ii) give different weights to the criteria selected (optional); and
  (iii) combine different criteria (if needed).

B (i)   select designated building sites;
  (ii)   evaluate the designated building sites; and
  (iii)   update the demand for housing remaining because of possible rejection of
          designated building sites in B (ii).

C (i)   based on the remaining demand (B (iii)), add possible building sites by way of on-
          screen digitizing, using some criteria as a background;
  (ii)   evaluate the possible building sites with respect to remaining criteria (which were
          not used as a background in C (i)); and
  (iii)   if needed, adjust the location, areal extent and form of the possible building sites
          (dependent on the outcome of C (ii)).

Steps C (i), (ii) and (iii) are part of a cyclic process, in which new building sites are
digitized and adjusted, until the total demand is satisfied. Some maps may illustrate the GIS
application. Figure 9.4 gives an example of a combination of elements from A and B (a
maximum distance of 5km from railway stations and greenbelts in which no building
activity is allowed). The two criteria combined here have been used to evaluate existing
building sites. In Table 9.1 the results of this evaluation are presented.

We found that more than one third of the areal extent of the designated building sites did
not satisfy the selected criteria. Therefore, and also because the number of dwellings in the
designated building sites would not be enough anyway, new building sites had to be
designed. Figure 9.5 gives an impression of the background against which on-screen
digitizing was carried out.

First, a part of the study area was selected (see also Figure 9.2). Some criteria have to be
indicated on the screen and possible building sites can be drawn. At a later stage, other
criteria will be selected and used for the final evaluation of the possible building sites. The
results can be presented graphically or in a tabular format. Figure 9.6 and Table 9.2
illustrate which parts of the possible building sites conflict with the nature criterion.

Table 9.1: Consequences for housing supply <5km from a railway station and not within a
            greenbelt for the period 1990-2015

Total areal extent of building sites in ha	8,580
Number of dwellings on building sites (estimated mean of 50 units per ha)	429,000
Total areal extent of building sites in ha inside zone 1	6,054
Number of dwellings to be built inside zone 1	302,700
Number of dwellings not to be built because of criterion 1	126,300
Total areal extent of building sites in ha outside zone 2	8,080
Number of dwellings to be built outside zone 2	404,000
Number of dwellings not to be built because of criterion 1	25,000
Total areal extent of building sites in ha inside zone 1 and outside zone 2	5,582
Number of dwellings to be built inside zone 1 and outside zone 2	279,100
Number of dwellings not to be built because of criteria 1 and 2	149,900

Notes: (i) Criterion 1: zone with distance to railway station that is <5km
       (ii) Criterion 2: zone not within a greenbelt

Figure 9.4: Areas outside greenbelts and within a certain distance (<5km) of railway stations

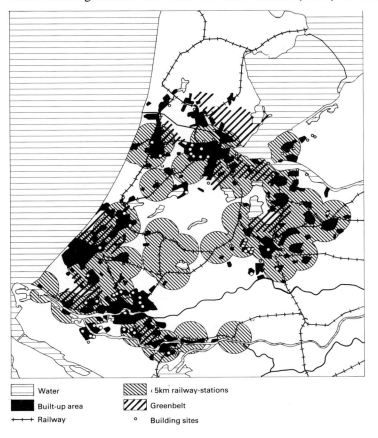

Water
Built-up area
Railway
< 5km railway-stations
Greenbelt
Building sites

Table 9.2: The amount of conflict between possible building sites and nature interests

Alternative building site (i)	Surface of design in ha	Amount of conflict in ha (ii)
A	444	15
B	550	145
C	510	135
D	2095	248
Total	3598	263

Notes: (i) See Figure 9.6 for the location of building sites
       (ii) Total areal extent of nature areas in grids (of 25 ha),
            even if they are only partially covered by an alternative building site

Figure 9.5: Alternative building sites in relation to the built-up areas and green belts

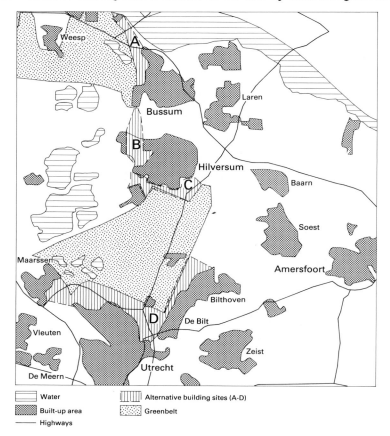

9.4   Ongoing research

In the feasibility study described above, the evaluation of existing building sites was easy. The main problem seemed to lie in the handling of existing databases needed for the evaluation. Several different databases with dissimilar data structures and scales were required in order to create map layers which could be used for the evaluation of building sites. Also, the design of new building sites by way of on-screen digitizing was difficult and the results were far from precise. In the next phase of the research project, scheduled to be completed in April 1990, the application will be made more user-friendly. A large number of databases will be used to define an extended set of criteria. All criteria will be made available to the users of the application in separate, easily accessible map layers. In this way planners should be able to select, combine and weigh the criteria they need for evaluating and designing building sites within a user-friendly environment. Also a special ARC/INFO macro will be developed to help planners design alternative building sites. Furthermore, a base map showing some general information and some basic criteria will be created in order to assist planners in the design process.

Figure 9.6: Alternative building sites in relation to nature areas

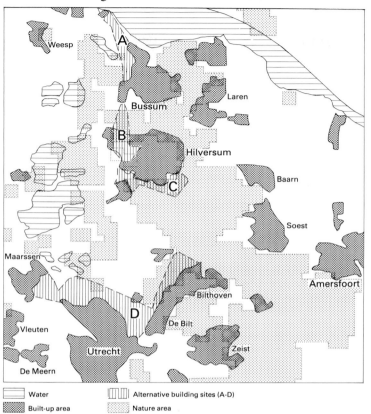

	Water		Alternative building sites (A-D)
	Built-up area		Nature area

9.5   Final remarks

Several elements of the GIS application have been described in some detail and were illustrated with maps and tabular output. A preliminary assessment of the pilot project indicates that it was a successful application of a geographical information system within a planning environment. Of course, the application has to be made more user-friendly; this will be done in the next phase by using the ARC/INFO Macro Language (AML) for programming the final GIS application. However, some critical remarks should be made. Firstly, for non-specialists (the second group of GIS users we referred to earlier) it is difficult to select and combine criteria and to give different weights to the criteria selected. This is even more difficult when some flexibility within this process is needed. Secondly, on-screen digitizing entails problems for non-specialists. The translation of the results of on-screen digitizing into real computer maps is time-consuming and contradicts the desire to interactively digitize, evaluate and adjust building sites according to the criteria selected. Thirdly, tabular information can be regarded as an essential part of the presentation of results in the GIS application. But a good tabular representation of the results of an evaluation of building sites requires extensive knowledge of the geographical information system used.

In summary, a common GIS like ARC/INFO is usually flexible, integral and efficient enough, but it is by no means interactive (user-friendly), although Arcshell offers a real improvement. Furthermore, it is only partially effective: the analytical functions are limited (integration with spatial models), the query function demands extensive knowledge of the GIS used and the on-screen design and evaluation of polygons (building sites for new houses) is cumbersome. This results in requests for the design and implementation - with the help of the Arc Macro Language - of specific, user-friendly and problem-oriented GIS applications. Concluding, we reiterate the importance of making GIS tools user-friendly and flexible because geographical information systems can be especially useful in decision-supporting environments like those described in this chapter.

References

Geertman, S.C.M. (1989) The application of a geographic information system in a policy environment: allocating more than one million dwellings in the Randstad Holland in the period 1990-2015, ARC/INFO Fourth Annual ESRI European User Conference, Rome, Italy

Geertman, S.C.M. and Toppen, F.J. (1989) Voorstel voor het Hoofdonderzoek Ruimte voor de Randstad, Ruimte voor de Randstad, deel 2, Faculty of Geographical Sciences, University of Utrecht, Utrecht

Harts, J.J. and Ottens, H.F.L. (1987) Geographic Information Systems in the Netherlands: application, research and development, in Proceedings of the International Geographic Information Systems (IGIS) Symposium: The Research Agenda, Volume III (Applications and Implementation), Arlington, Virginia, USA, pp. 445-457 (November 15-18)

Jong, T. de and Ritsema van Eck, J. (1989) GIS as a tool for human geographers: recent developments in the Netherlands, Paper presented at the Deltamap User Conference, Fort Collins, USA, (April 11)

Ministry of Housing, Physical Planning and Environment (1988) On the Road to 2015, Comprehensive summary of the Fourth Report on Physical Planning in the Netherlands, SDU Publishers, The Hague

Ministerie van Volkshuisvesting en Ruimtelijke Ordening (1988), Vierde Nota over de Ruimtelijke Ordening, deel a: beleidsvoornemen, Tweede Kamer, vergaderjaar 1987-1988, 20.490, (1 en 2), SDU, 's-Gravenhage

Scholten, H.J. and Padding, P. (1988) Working with GIS in a policy environment, ARC/INFO Third Annual ESRI European User Conference, Kranzberg, W.Germany

Toppen, F.J. and Geertman, S.C.M. (1988) Ruimte voor een Miljoen Woningen in de Randstad?, Ruimte voor de Randstad, deel 1, Faculty of Geographical Sciences, University of Utrecht, Utrecht

Stan C.M. Geertman and Fred J. Toppen
Rijksuniversiteit Utrecht
Geografisch Instituut
Postbus 80115
3508 TC Utrecht
The Netherlands

10 GEOGRAPHICAL INFORMATION SYSTEM APPLICATIONS IN ENVIRONMENTAL IMPACT ASSESSMENT

Jörg Schaller

10.1 Introduction

Geographic Information Systems (GIS) have become of increasing significance in recent years. One main reason for this in the field of Environmental Impact Assessment (EIA) is the need to compare a great number of area-related data describing the natural resources involved and their sensitivity to the effects of various impacts. Because GIS can be used to couple area-related data with their attributes, and can be used to overlay these, they represent highly efficient instruments for such planning tasks.

This chapter describes the basic structure of area-related databases and the use of GIS in selected examples of applications from recent EIA projects. The four examples are as follows;

(i)   EIA study for the channel of the River Danube between Straubing and Vilshofen;
(ii) ecological balancing in land consolidation;
(iii) determination of noise pollution effects on residents near the new Munich II airport;
(iv) EIA for the Federal A94 motorway.

Each of these chosen research and planning projects has involved the ARC/INFO software package created by Environmental Systems Research Institute (ESRI). For each project a different GIS database was created and evaluated by the ESRI-Germany application group.

10.2 EIA study on the River Danube

The aim of this study was to investigate possible effects of different river works on the natural resources of the Danube and its floodplain. A null hypothesis was also considered. Potential risks and deleterious effects have been identified over time to bring ecological considerations to the fore before the building works begin. The study area is a section of the Danube between Straubing and Vilshofen, about 50 km long with an average width of about 5 km. The total area covered is 257 sq. km.

Conception

This ecological study, which is to be finished in 1989, has been undertaken within a conceptual framework that has been developed in co-ordination with the Bavarian State Agency for Environmental Conservation. This concept breaks up the project into five main phases (Figure 10.1) which in part run simultaneously (Koeppel et al. 1988, Planungsburo Dr Schaller 1988).

Data collection

After the initial project definition phase with the client and the Bavarian State Agency for Environmental Protection and Conservation, representatives of several disciplines from different sources (consulting offices, universities and own resources) were invited to a workshop where

H. J. Scholten and J. C. H. Stillwell (eds.),
Geographical Information Systems for Urban and Regional Planning, 107–117.

the main issues, the data needs and principles of assessment were discussed. A work programme was developed in which all questions of data collection and processing were considered.

Figure 10.1: Project phases

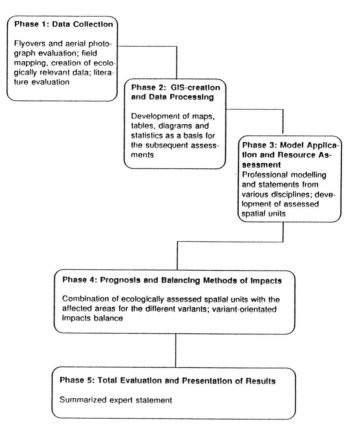

GIS structure and data processing

The area-related base information of the GIS that was constructed is illustrated in Figure 10.2. For the entire study area, a map library of eighty one maps (scale 1:5,000) was prepared, containing all necessary information for the area-related Environmental Impact Assessment. The data were either taken up directly in digital form, or manually digitized using aerial photographs, interpretation maps and resources maps showing field boundaries. The corresponding attributes were stored in the INFO data bank. Statistical analysis of basic data has been carried out parallel to data processing, e.g. evaluation of groundwater time series data, to create results which have been assigned as single attributes to the GIS database.

Figure 10.2: Area-related base information of the GIS

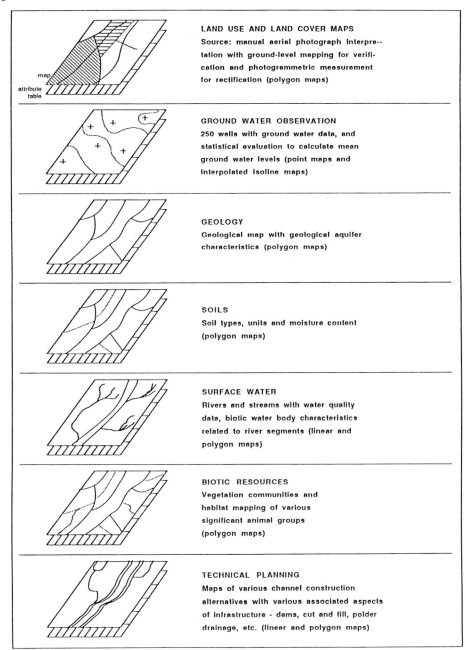

LAND USE AND LAND COVER MAPS
Source: manual aerial photograph interpre--
tation with ground-level mapping for verification and photogrammetric measurement
for rectification (polygon maps)

GROUND WATER OBSERVATION
250 wells with ground water data, and
statistical evaluation to calculate mean
ground water levels (point maps and
interpolated isoline maps)

GEOLOGY
Geological map with geological aquifer
characteristics (polygon maps)

SOILS
Soil types, units and moisture content
(polygon maps)

SURFACE WATER
Rivers and streams with water quality
data, biotic water body characteristics
related to river segments (linear and
polygon maps)

BIOTIC RESOURCES
Vegetation communities and
habitat mapping of various
significant animal groups
(polygon maps)

TECHNICAL PLANNING
Maps of various channel construction
alternatives with various associated aspects
of infrastructure - dams, cut and fill, polder
drainage, etc. (linear and polygon maps)

A process is followed to integrate the different single GIS databases required to conduct the area-related assessments (Figure 10.3). Integrated resource assessment maps are created by using GIS procedures such as overlay, buffer and network techniques. The different construction alternatives become subsequently overlaid with the resource data, resulting in geographic and statistical information for interpretation in the evaluation of those construction alternatives.

Figure 10.3: Integration of the GIS database for the EIA

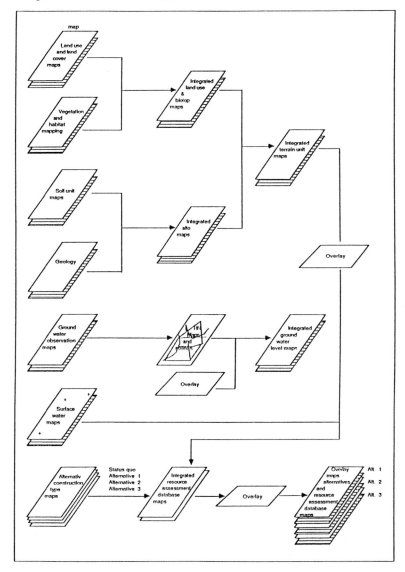

Model application

In addition to more or less static evaluations, the dynamic effects of the respective impacts on ground and surface water conditions must also be considered (e.g. groundwater level changes, changes in water quality, changes of vegetation over time, etc.). In order to analyse groundwater conditions for example, a two-dimensional stationary groundwater model was used, coupled with the GIS database (Furst et al. 1989). The basis for the model calculation is a grid with a mesh width of 500x250 metres. For every grid element, aquifer parameters (kF value, dam or gravel overlaid, etc.) and hydrological parameters (groundwater table, new formations thereof, etc.) are coupled with other resource data (e.g. vegetation, soils) and valued. Similarly, the water quality conditions before and after the works are compared for each alternative. As an example of a biotic evaluation in the form of a spatial model, the GIS application for the derivation of values of the biotope natural should be mentioned. Calculation procedures from the biological sciences were used to compute a 'stepping stone' function of different habitats for various animal and plant species (Schreiner 1989). The following working steps of 'stepping stone' assessment were carried out:

(i)   selection of polygons which have to be ascertained (e.g. special biotope types, surface water, dry meadows, hedgerows, etc.);
(ii)   triangulation of polygon borders (Z-values of the TIN cover identify the biotope areas);
(iii) selection after biotope 'ID' and length of connections (connection lines are only stored when they contain different biotope IDs and when they are shorter than the allowed maximum distance - optional as programme ground);
(iv)   selection of the shortest distance between two similiar biotope IDs;
(v)   statistics on the number of connection lines of each biotope, defining a 'connection value loss'; and
(vi)   assigning the connection value class as 'stepping stone' value to the attributes of the initial polygons.

Figure 10.4 shows the triangulation of the chosen biotopes from the data content of the land use-land cover map produced using the TIN software package. The range of the shortest respective space between the biotopes is also shown (Manegold 1989). In the project, the various spatial valuation techniques were applied according to disciplinary requirements. Thereby it was possible to differentiate between the dependencies of the database and the evaluation methods. All model and evaluation results were treated in an area-related fashion in the GIS however, and represent an integrated total balance for further analysis.

Prognosis and evaluation of impacts

The study considered the following impacts:

(i)   the status quo - null hypothesis;
(ii)   effects of construction;
(iii) effects from the form of the works (variant comparison); and
(iv)   effects from channel river operation (wave action).

The fundamental procedure is illustrated in Figure 10.5. Depending on the type of valuation, the various effects are described in the form of transition matrices (e.g. land use changes) or risk profiles. In this way, it is possible to compare the effects of impacts both for the status quo as well as the variants with one another, and thereby develop an 'ideal' model (Koeppel et al. 1988, Planungsburo Dr Schaller 1988, Schaller 1989).

Figure 10.4: TIN procedure for the 'stepping stone' evaluation

TRIANGULATION OF
POLYGON BORDERS

SELECTION AFTER LENGTH
OF CONNECTIONS

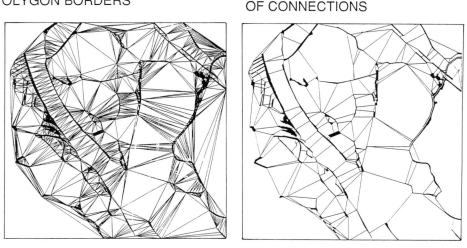

Source: ESRI - J. MANEGOLD [7]

10.3 Ecological balancing in land consolidation

The increasing ecological awareness of recent years has had important effects on land consolidation. In addition to the primary economic questions, ecological aspects such as biotope protection, resource protection and endurance of agricultural production by minimalization of environmental impacts are all current issues in modern land consolidation. Methods for ecological balancing in land consolidation were developed in a research project at the University of Technology, München-Weihenstephan, commissioned by the Bavarian Ministry of Agriculture (Auweck et al. 1989, Bachhuber and Schaller 1988). In addition to dealing with fundamental questions of resource evaluation, geographic information systems were also used for data coupling. The GIS database is calculated from the inventory of the natural resources and the planned land use changes. As well as the classic evaluation parameters such as groundwater quality and quantity, surface water bodies, soil and soil erosion, local climate and effects on vegetation and fauna, integrated ecological parameters such as diversity, habitat networks and edge effects must also be considered. An analysis of changes to the quality of landscape elements over time is likewise essential.

Maps can be generated which show the current status of the allotment division in the study area of Oberhaselbach, and as an example, a planning variant of the spatial use after consolidation. The changes from clear-cutting, new tracks, slope direction, distance of small landscape structures etc. allow the storage and blending of the base geometry, the setting-up of difference statistics, all related to areal size, clear-cutting length, edge effects (e.g. neighbouring types) and structural losses. Plate 10.1 depicts the effects of land use change by land consolidation on the soil erosion. Using a TIN cascading procedure, the direction of surface water flow and the watersheds were calculated. By using the soil loss equation the

Figure 10.5: The evaluation procedure

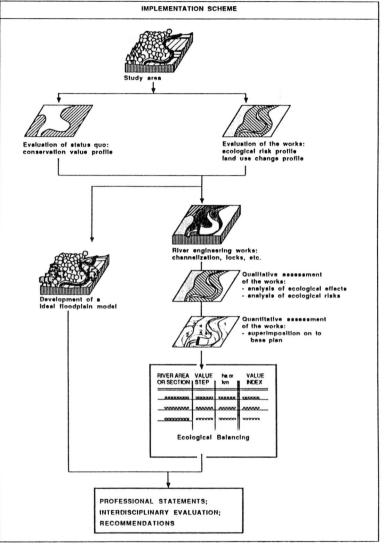

Source: J.G. KÖPPEL, Planning Office Dr. Schaller

quality of soil losses within the watershed can be calculated for every alternative strategy of consolidation. Using GIS evaluation, it is possible to compare very quickly the different variants for track and land consolidation planning. By examination of the habitat network, it is possible to quantify and illustrate the existing network and thinning out effects. The coupling of clear-cutting lengths, slope, soil types and uses allow area-related calculations of erosion and deposition of pollutants in river bodies. With the aid of the technical evaluation basis and the possibility of GIS input, land consolidation planning has now completely new instruments at its disposal to support the necessary technical and planning decisions (Auweck et al. 1989, Bachhuber and Schaller 1988, Haber and Schaller 1988).

10.4 Determination of noise pollution effects on residents near the Munich II Airport

With the publication of the initial flightpath proposals for the new Munich II airport came sharp public protest and questions as to who could be affected in the vicinity. Geographic information systems, coupled with noise dispersion calculations, have served to determine the area-related effects on local communities, and to illustrate these statistically and cartographically. Through overlay of the calculated noise values on the ground (from standard aircraft noise calculation procedures) with the population size of the communities, it is possible to calculate quickly who will be affected by sporadic or continuous disturbance. This can be related to airport management models, aircraft types and chosen flightpaths. Figure 10.6 shows, as an example, a three-dimensional representation of affected communities in the vicinity of the airport, whereby for a particular flightplan variant and airport management model, the population sizes are weighted against dBA noise level values and are illustrated as 'annoyance mountains'. Such GIS evaluations are very effective, and can be easily conducted and presented for lay people (ESRI 1988).

Figure 10.6: Three dimensional view of noise levels and annoyance of the population

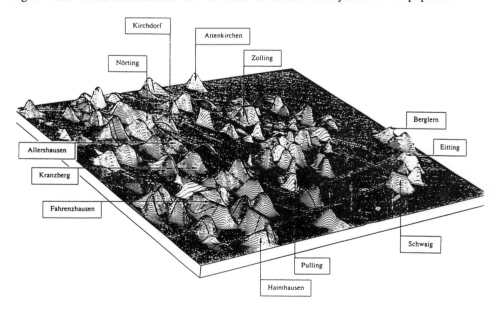

10.5 EIA study for the Federal Motorway (A94)

An EIA study has been carried out for the Autobahn Construction Authority to compare two different routing alternatives with several planning variants of autobahn construction between Munich (Ampfing) and Passau (Forstinning) in Southern Bavaria. GIS capabilities have been used for the assessment procedures and to produce the results in the form of statistical analysis and cartographic displays. By using conventional methods in data analysis, advanced GIS tools of data processing such as overlay procedures, TIN and NETWORK applications led to sophisticated assessment methods for a large data base. Four types of geographic data have been evaluated:

(i)   abiotic resource data;
(ii)   biotic resource data;
(iii)   land use and impact data; and
(iv)   planning data of the line variants.

A data bank of nine maps (scale 1:10,000) was developed and an integrated evaluation was carried out thereon. The assessment of impacts and eventual motorway operation were considered for three evaluation areas:

(i)   landscape ecological terrain units;
(ii)   vegetation valuation; and
(iii)   fauna valuation.

The ecological valuations and their criteria are illustrated in Figure 10.7.

For a comparison of variants, the routes were digitized and blended with the base data, with a 200 metre buffer zone from each side of the central line, which has been differentiated by form of construction (levels, dammed, cut, etc.). Figure 10.8 depicts the map with the assessed ecological terrain units by five levels of sensitivity overlaid by the 400 metre buffers of the construction variants. Through a comparative interpretation of the cartographically and statistically determined 'affected aspects of the natural resources', it was possible to distinguish between optimal and most unfavourable variants. The planning corridors could thus be valued in relation to their ecological sensitivity (Schaller 1987).

10.6 Conclusions

The chosen examples of environmental impact assessments have been carried out using GIS methods and procedures. Principally this planning work could also be done by conventional methods but the drawback with this is the very time consuming nature of the manual work involved. Environmental impact studies use very large databases of natural resource and land use data and normally different alternatives of impacts have to be assessed. Additionally, the results of several alternative impacts on the natural resources have to modelled using different methods such as distribution models, pollution models, spatial effects of impacts, changes of land use etc. Therefore GIS, in connection with impact models, are very useful tools to carry out such sophisticated work, because after having automated the resource databases, changes in land use or types of impacts can be modelled easily and quickly and results can be presented that are of a very high standard of quality.

Figure 10.7: Ecological assessment criteria

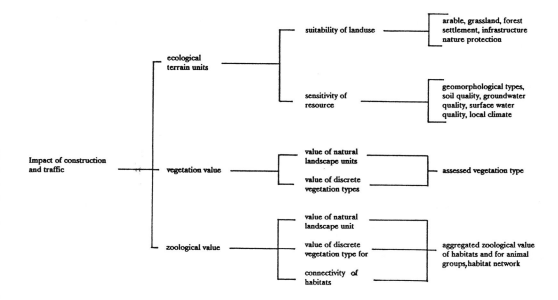

Figure 10.8: Ecological assessment of alternative routes of Federal motorway construction

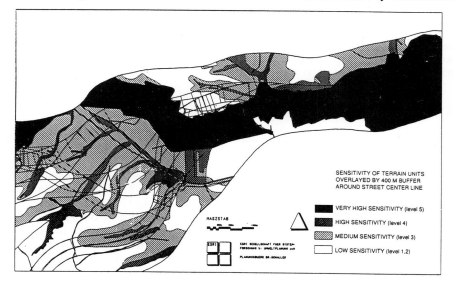

References

Auweck, F., Bachhuber, R., Riedel, A., Theurer, R. and Schaller, J. (1989) Handbuch zur Ökologischen Bilanzierung in der Flurbereingiung, Abschlußbericht, Lehrstuhl für Landschaftsökologie TU München-Wiehenstephan, (unpublished)

Bachhuber, R. and Schaller, J. (1988) Mehtoden zur Umweltverträglichkeitsprüfung (UVP) agrarstruktureller Veränderungen, Schriftenreihe Bayer, Landesamt für Umweltschutz, München, 84, 65-81

ESRI (1988) Lärmgutachten zu den neuen Flugrouten Flughafen München II, Auszug: Lärmzonenberechnung nach AzB, Erstellung im Auftrag der Gemeinde Kranzberg, (unpublished)

Furst, J., Nachtnebel, H.P. and Remmel, I. (1989) Ökologische Rahmenuntersuchung Straubing-Vilshofen/Grundwassermodell, Vorbericht Variantenstudie, Universität für Bodenkultur, Wien, (unpublished)

Haber, W. and Schaller, J. (1988) Connectivity in Landscape Ecology, Proceedings of the 2nd International Seminar of the 'International Association for Landscape Ecology', Hrsg.: K.F. Schreiber, Münstersche Geographische Arbeiten, 29, 181-190

Koeppel, J.G., Mayer, F., Schaller, J. and Steib, W. (1988) Konzept der Ökologischen Rahmenuntersuchung zum geplanten Donauausbau zwischen Straubing und Vilshofen (BRD), Wissenschaftl, Kurzreferat zur Arbeitstagung der IAD in Mamaia/Rumänien, (in preparation)

Manegold, J. (1989) NETNET, Programm zur Trittsteinbewertung, ESRI, Kranzberg, unveröffentlicht

Planungsburo Dr Schaller (1988) Neue Schritte beim Donauausbau - Die ökologische Rahmenuntersuchung, In RMD-Intern 3/88 und 1/89, Rhein-Main-Donau AG (Hrsg.), München

Schaller, J. (1987) Environmental Impact Assessment (EIA) of different routing alternatives of a planned autobahn between Munich and Passau (Southern Bavaria), In the Proceedings ARC/INFO User Conference 1987, ESRI Redlands, California

Schaller, J. (1989) Environmental Impact Assessment Study for the planned Rhine-Main-Danube River Channel (Federal Republic of Germany), in Proceedings ARC/INFO User Conference 1989, ESRI Redlands, California

Schreiner, J. (1989) Fachliche Vorgaben zur Trittsteinbewertung, In Zwischenbericht der Projektleitung, In Planungsbüro Dr. Schaller, Arbeitsgruppe Ornithologische Arbeitsgemeinschaft Ostbayern, Projektbericht 'Ökologische Rahmenuntersuchung zum geplanten Donauausbau zwischen Straubing und Vilshofen - Bewertungsprogramm', (unpublished)

Jörg Schaller
Environmental Systems Research Institute (ESRI)
Gesellschaft fur Systemforschung und Umwelltplanung mbh
Ringstrasse 7
D-8051 Kranzberg
Germany

11 A GEOGRAPHICAL INFORMATION SYSTEM BASED DECISION SUPPORT SYSTEM FOR ENVIRONMENTAL ZONING

Hans E. ten Velden and Gerald Kreuwel

11.1 Introduction

A GIS based system is being constructed to support the decision making process related to environmental zoning policy and its enforcement in the Netherlands. The aim of environmental zoning policy is to minimize the environmental effects of industrial noise, odour nuisance and the risks resulting from the dispersal of toxic compounds and from the danger of explosion. Nowadays this kind of integral policy making is one of the important issues in land use planning and environmental policy. This chapter describes the general layout and ideas underlying a support system which is in the process of being developed. After an explanation of some of the rationale and context for environmental zoning policy in the Netherlands has been provided, the main functions of the application will be outlined. Although there are indications that a successful implementation of the system will depend equally on the organizational aspects of the system as on technical know-how, the focus here is on the latter, relating to the description of the problem. The specific problems associated with the implementation of relatively small and dedicated systems in local or provincial public service organizations requires a different approach.

The technical challenge has been to design a very 'open' and flexible system in order to be able to use existing digital topographical and environmental quality data. Furthermore, it is necessary to be able to adapt to different user requests during implementation. By bringing together geoprocessing software and environmental transmission models, the application can be described in terms of a decision support system (see Section 11.3). The integration of technical knowledge about environmental pollution and systematic information on land use and environmental quality into one computerized system will provide a powerful tool for decision making. Rational decision making will be supported by the rapid evaluation of changes in pollution levels and land use in terms of environmental quality.

11.2 Environmental policy and land use planning

Policy context

Studies of the causes and effects of soil, water, and air pollution and of noise nuisance and risk reveal increasingly that land use planning should play an important role in the prevention of pollution and in the development of emission reduction programs for polluted areas. Almost every pollution problem has both a spatial and a temporal dimension. Therefore long term strategic land use planning, taking into account the prevention and solution of environmental problems, is essential. The integration and alignment of land use planning and environmental policy is now taking place in several ways. One of these is through environmental zoning - creating physical, jurisdictional and/or attention zones around emission sources like industrial areas or roads. Physical zones can be defined as those that contain no sensitive functions, whereas jurisdictional zones are those in which the measured level of actual of future pollution levels permits only limited or no additional housing

H. J. Scholten and J. C. H. Stillwell (eds.),
Geographical Information Systems for Urban and Regionai Planning, 119–128.
© 1990 *Kluwer Academic Publishers. Printed in the Netherlands.*

construction. Attention zones are zones in which a detailed survey of pollution levels is required in advance of house construction. This type of zoning is in fact an old planning concept, used in the 17th century. Different forms of building restrictions in these areas occur as a consequence. This kind of policy, aimed at minimizing the effects of environmental problems, is often contradictory to the aims of efficient land use. Problems associated with adjusting both of the policy fields involved derive from the fact that land use planning in the Netherlands draws upon a long tradition of plan making. Systematic environmental policy and planning however are still under construction. This is true especially for the development of legislation relating to emission norms. The close relationship between land use and environmental planning is most pronounced at the local and regional scale. The most intensive decision making on land use takes place at these levels. In general, the desired combination of policies relating to land use and environmental quality depends upon the degree to which causes and effects of certain types pollution can be related to the functional use of land and water. The causes may be a particular configuration of industrial areas or a road network. The effects can be related to the degree to which sensitive functions are exposed to the pollution.

In the Netherlands, environmental zoning is becoming a more integral component of strategic regional planning and decision making. For the preparation of these plans, both land use (zoning included) and emission reduction scenarios have to be drawn up for the municipal or provincial government to make decisions about. To support the preparation of policy proposals and presentation to a general public, a GIS based system can prove very helpful.

Environmental zoning in the Netherlands

In the case of environmental zoning, causes and effects of pollution (e.g. nuisance and risk) are clearly related to functional land use. Zoning is a planning concept in which sources of pollution are physically or spatially separated from new land use functions that are sensitive to specific sorts of nuisance and risk. The example of safety zones round nuclear plants is well known. In the Netherlands, the government has introduced the principle of double norms. Emission norms relate to individual sources whereas immission norms relate to specific land use functions and soil types. One consequence of these norms is the existence of different forms of building restrictions for sensitive functions in areas with a certain level of nuisance. High levels of pollution or risk relative to the norm may result in houses being demolished. The land use functions that are sensitive in this respect include housing, recreation, hospitals, schools and public space. One characteristic of forms of pollution that necessitate zoning is that their influence diminishes with distance. That holds for noise, odour, risk of explosion and the dispersal of toxic compounds from stationary sources. The question of how to deal with a particular combination of different types of nuisance and risk in one area is now under consideration.

Because land is scarce in the Netherlands, the Dutch devote a great deal of time and money to attempts to optimize land use. In order to contribute to this goal, a permanent effort needs to be made to minimize the quantity of land designated for environmental zoning. This policy is officially declared in the Dutch national plan for environmental protection (VROM 1989). In densely populated areas like the Randstad, the enforcement of environmental zoning regulations often causes land use planning problems as in the case of large airports like Schiphol, for example. This also occurs in areas where there is great pressure on more intensive use of land for housing and amenities, and in newly developed areas. Optimization in these situations involves generating a balance between measures to be taken at the source (reducing emissions) and measures to reduce the effects of pollution, by environmental zoning, for instance.

The GIS based system that is now being developed is devoted to this balancing process. It is needed to support both long term and short term decision making, where long term decisions are closely related to strategic land use planning. For long term decision making it should be possible to compute land use effects of different types of emission reduction policies. Short term decision making, on the other hand, is more related to the need for quick response in situations where a judgement has to be made about whether or not a permit for additional emission should be granted.

11.3 General structure of a decision support system

The design of the system can be described as a Decision Support System (DSS). In short a DSS contains a combination of databases, computational models, analysis and evaluation modules and a user interface (Figure 11.1).

Figure 11.1: General structure of a decision support system

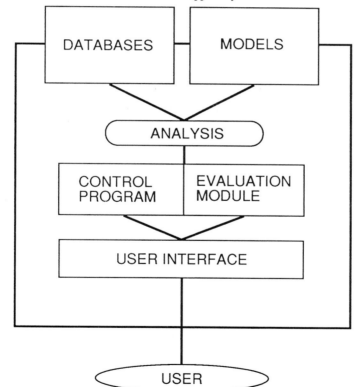

The characteristics of this general framework have been well documented in the literature on decision support systems (see Van der Vlugt 1989, Sprague and Carlson 1982, Van der Heijden 1986, Scholten and Van der Vlugt 1988, Ten Velden and Scholten, 1988, Voogd 1983, and Fedra and Reitsma 1989). Decision support systems:

(i)   are computer systems;
(ii)   are dedicated to a restricted but well defined area of application;
(iii)   are systems with computational modelling and analysis techniques and with data storage, retrieval and evaluation techniques;
(iv)   do not 'take' decisions but facilitate the logistics of the decision making process; the system does not produce the answer to a problem but provides an arranged selection of alternative solutions;
(v)   are interactive systems that help the decision maker by posing questions systematically;
(vi)   produce custom-built information; and
(vii)   should contain a user friendly (graphical) interface and should have very short response times.

Environmental zoning as a planning problem fits conveniently into the general structure of a DSS. Numerous databases (attribute layers) are created and have to be internally administered. There are at least three different transmission models involved (for noise, risk and dispersal of toxic compounds through the air), and finally there is a need to analyse and evaluate different land use planning and emission reduction scenarios.

The aim of environmental zoning is the continuous maximization of environmental quality and minimization of restrictions on land use. The system can be used to judge the effects of certain policy measures and/or alternative land use plans. Through integrating spatial information and sets of model outputs, it will also be possible to search systematically for optimal locations for emitting sources and sensitive functions. The next section presents an outline of the different modules of the system.

11.4 Functionality of the system

Main functions

The functionality of the system can be divided into six main elements from which A, C, D and E are directly related to the use of GIS software:

(A)   data entry, storage, retrieval and management;
(B)   computation of environmental quality with transmission models;
(C)   spatial (land use) and environmental analysis;
(D)   comparison and evaluation of alternatives;
(E)   presentation and mapping; and
(F)   user interface and scenario generator.

The general layout of the system is illustrated in Figure 11.2.

In the following section, we will focus on those aspects that are related to the use of GIS software. Transmission models (B) are discussed as far as GIS compatibility is concerned. The scenario generator is a specially designed piece of software. It is closely related to the user interface as it should enable the user to compose emission reduction scenarios, for example, interactively. This requires development of user friendly and intelligent interfaces for model input. On the other hand, site allocation scenarios for land use sensitive to nuisance and risk could be drawn up interactively, constrained by environmental criteria. Minimizing the number of houses affected by a certain level of nuisance while maximizing proximity to public transport is an example of the latter.

Figure 11.2: Structure of the application

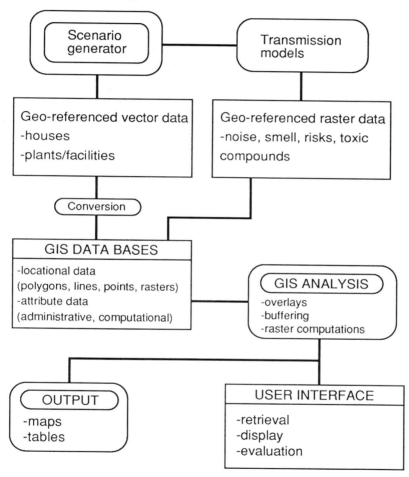

Integrated processing of vector and raster data

The problem to be solved here has involved the integration of processing two different types of data. In GIS software there are now three main types of data notation: vector format, straight raster format and quadtree format. The quadtree data format is relatively new to GIS software. It is based on maximum raster size as a function of data homogeneity. The advantage is fewer rasters and this results in more efficient storage and processing as a consequence. In the application to environmental zoning, both straight raster and vector data are processed: administrative and empirical data on land use (attributed to vectors) and measured or computed data on nuisance and risk levels (attributed to rasters). The former type contains locational data on sensitive objects like houses, amenities, recreational areas as well as the sources of pollution. The latter are raster data, output from the transmission models.

Most Dutch municipalities do not yet have a complete digital locational database. That is why it was decided for the time being to work with a governmental topographical database (De Jong 1989) at the 1 to 25,000 scale (containing the type of land use as attribute information), for interfacing and presentation purposes, enriched with a very accurate cadastral database (scale 1 to 500) containing the exact location of buildings with some extra attributes like type of use, age and address. This database will be enriched further for analysis purposes. The topographical database must be organized in different layers, a feature characteristic of GIS data structure. It facilitates greater flexibility in presentation and output. It is also possible to add new layers, such as data on soil pollution which happens to be a problem in a lot of Dutch municipalities because of the restrictions it imposes on building. This information seems to have become vital in local land use planning and is therefore relevant to environmental zoning policies.

The raster format of the model output data is necessary to compute a desired form of aggregation of different sorts of nuisance and risk. Overlay of raster data can be seen as the computation of matrices. For an accurate interpolation of raster values at the edge of the contour, the relevant raster cells are divided into smaller segments. This provides a more accurate localisation of objects (Figure 11.3). For nice looking presentation and mapping, the rasters are automatically converted into vectors by interpolation of the raster cells. Because at least three different transmission models are used, computations are made for each individual industrial area and alternatives are produced, the database has to be organized properly. In setting up and using the system, a large number of maps and layers will be produced. Therefore a standard map layout with sufficient information is recommended. Good admistration and labelling will prove to be essential for reliable and flexible use. The power of the software is in the layered organization of the topographical data and a well administered database. It facilitates, for example, the selection of individual industrial areas with their spatial spheres of influence for more detailed analysis.

Transmission models and GIS

The combination of transmission models and GIS makes it possible to generate alternative solutions for the process of decision making on environmental quality management. Technical emission reduction scenarios can be drawn up, processed by transmission models and analysed using GIS functions on both raster and vector levels. Transmission models are mathematical algorithms based on formulae supplied by the government. Nevertheless, the interpretation of these rules has led to differences. Handling of the transmission models is skilled work. The input sets of the models demand estimation of source capacities and insights into the screening function of buildings (i.e. the degree to which buildings act as a shield against pollution). Of course it is possible to develop intelligent interfaces with these models, i.e. to produce a comprehensive scenario generator. This, however, is a long term perspective.

Figure 11.3: Refined raster for accurate interpolation and localisation

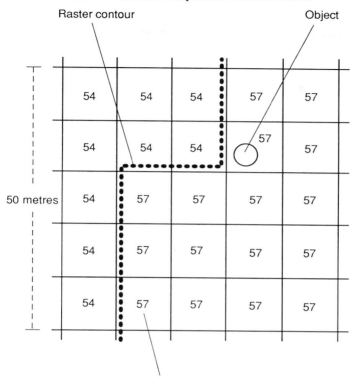

Computed values in data base

Environmental zoning analysis

Analysis to be undertaken during the decision making process of environmental zoning can be divided into inventory and decision-oriented. The process begins with an inventory of emissions and immissions. Emissions are levels of pollution at sources whereas immissions refer to levels of pollution at particular destination localities. This situation can be mapped and analysed for its land use implications. The criteria may be the number of houses affected, and/or the proportion of recreational and other sensitive areas affected which may include the locations of functions like hospitals, schools etc. The number of houses affected by certain levels of (industrial) noise is the decisive measure for legislative purposes. However, if a good balance of measures is to be achieved between emission reduction on the one hand and zoning measures on the other, it is important to decide on which spatial elements should be included in the balancing process besides housing. The more building restrictions are introduced, for example, based on some form of aggregation of nuisance and risk factors, the more important a precise inventory of effects is. As in most environmental problems, economic interests tend to play an important role. For planners it is a challenge to translate zoning restrictions into real costs.

In the decision making phase, desired future land uses and spatial structures have to be input by strategic land use planners. This process includes the specification of both locations of sensitive functions and location of sources of emission. Integration of a desired spatial pattern of immission levels then generates constraints for a search in emission reduction scenarios. On the other hand the future spatial pattern of emissions generates constraints for the locations of sensitive functions. Once the municipal government has decided on a spatial structure plan including environmental zones, the system can be used as a tool for policy maintenance.

The system should generate the following type of information for analysis purposes:

(i)   the number of dwellings within a specified zone or number of zones, subdivided into age classes;
(ii)   the proportions of specific functional areas under a certain zone;
(iii) lists of industries with a certain level and type of emission;
(iv)   lists of locations with a certain nuisance level;

The level of information depends of course on the richness of the land use database. If it is being used for drawing up a scenario, the system should be able to answer 'what if..?' questions. For example:

(i)   what level of emission reduction is required from which source if immission levels in a certain area were to fall by 50 percent?
(ii)   what costs are involved if that measure is taken?
(iii) what effect does an additional source of noise have on the zone for noise nuisance?

The system supplies a range of relatively simple but effective techniques to support these kinds of questions. Mathematical overlays can produce maps in which the spatial effects occurring when different sorts of nuisance and risk are aggregated, are represented. Simulations can be carried out by overlaying contours and topography (enriched with attributes) on the basis of possible extra emissions or emission reductions. The addition of a cost analysis module would provide the capability of a financial comparison of the costs of emission reduction and costs of land use restrictions. This would be possible using raster processing techniques. There is at the moment, however, no accepted technique for estimating the latter type of costs.

Presentation and mapping

An important product of the system that has just been described is the hard copy output of maps and tables. Plate 11.1 illustrates, using an example of the Dutch municipality of Maastricht, how noise contours can be overlaid on maps of residential areas. Those residential areas affected by different levels of noise pollution can then be identified and graphically presented. In a decision making process, communication between politicians, policy makers and civilians will be improved with the availability of maps. The function of visualization especially in a GIS context is described by Ten Velden and Van Lingen (1989).

11.5 The user interface

Special attention is given to the user interface because the potential user does not have specific knowledge of GIS software. Besides that, the basic software used is not dedicated to a specific application. Most commercial GIS software packages contain a relatively poor user interface. Unlike the extremely user-oriented interfaces of Macintosh-like software, very often mainframe and PC-based GIS software is lacking easy-to-adapt menu driven user interfaces. Although some progress is being made at the moment for the application we have described, a new user interface shell has had to be developed. Especially in a geoprocessing environment, visualization of systems operation and data processing is very important (Ten Velden and Van Lingen 1989). Users of the system are not GIS experts at all. Most are not familiar with command languages whatsoever. The basic menus provided with the package that we have used (Geopakket) were much too general. In such circumstances, there are two possible courses of action. Either users of the application have to be educated in the use of the software package, or it becomes necessary to develop a application-specific user interface. The latter option was chosen in this case. To maintain flexibility in the system, a so called 'menu generator' was built with which any sequence of operations could be generated very easily. This piece of software generates the command code for the basic GIS software.

It is important to stress that a successful implementation and integration of decision support systems depends on the time needed to learn to work with the system. A well designed user interface will shorten learning time dramatically.

11.6 Concluding remarks

This GIS application for environmental zoning will be implemented in three Dutch municipalities. On the basis of these experiments the system has to prove its value in supporting the decision making process. The fact is that the use of GIS based systems facilitates not only the enforcement of norm-based environmental policy but also norm-based land use planning. More generally, the attachment of environmental immission norms to land use databases will support a more systematic environmental bias in decision making on emission reduction and/or land use planning. Much will depend on how concerned the local government is with environmental (zoning) policy.

Technically the integration of vector and raster type data within one system is an important step forward in geoprocessing techniques. Vector data enable more accurate mapping whilst raster data allow easier and more rapid calculations to be made. The vector-raster integration also facilitates the inclusion of photographs and documents as attributes of points or polygons. The search for usable spatial analysis techniques that support decision makers in the field of land use planning and environmental policy is made easier by using the computer. The scientific structuring of this relationship more generally awaits further development.

References

De Jong, W.M. (1989) GIS database design: experiences of the Dutch National Physical Planning Agency, Paper presented at the GIS Summer Institute, Amsterdam, (14-25 August)

Fedra, K. and Reitsma, R. (1989) Decision support and GIS, Paper prepared for the GIS Summer Institute, Amsterdam, (August 14-25)

Heijden, R.E.C.M. van der (1986) A decision support system for the planning of retail facilities, Technische Universiteit, Eindhoven

Scholten, H.J. and Vlugt, M. van der (1988) Applications of Geographical Information
    Systems in Europe, a state of the art, Paper prepared for the Unisys Management
    Seminar on GIS, Sydney, (October 25-26)
Sprague, R.H. and Carlson, E.D. (1982) Building Effective Decision Support Systems,
    Prentice Hall, Englewood Cliffs
Velden, H.E. ten and Lingen, M. van (1989) Visualization and GIS, Paper prepared for the
    GIS Summer Institute, Amsterdam, (August 14-25)
Velden, H.E. ten and Scholten, H.J. (1988) De betekenis van DSS en ES in de ruimtelijke
    planning, Bouwstenen 11, Tecnische Universiteit, Eindhoven
Vlugt, M. van der (1989) The use of a GIS based DSS in physical planning, Paper
    prepared for the GIS/LIS Conference, Orlando, (November)
Voogd, H. (1983) Decision support systemen voor overheidsplanning?, Enkele introduceren-
    de kanttekeningen, Planologisch Memorandum 1983-4, Technische Universiteit, Delft
VROM (1989) Kiezen of Verliezen, First National Policy Plan for Environmental Protection

Hans E. ten Velden
Rijksplanologische Dienst
Willem Witsenplein 6
2596 BK Den Haag
The Netherlands

Gerald Kreuwel
GEOPS B.V.
Vadaring 2
6702 EA Wageningen
The Netherlands

12   MULTICRITERIA ANALYSIS AND GEOGRAPHICAL INFORMATION SYSTEMS: AN APPLICATION TO AGRICULTURAL LAND USE IN THE NETHERLANDS

Ron Janssen and Piet Rietveld

12.1 Introduction

Regional and urban planning problems are often hard to solve for various reasons. Large amounts of data are needed, due to the number of spatial units involved or the range of phenomena taken into account. Uncertainties of various kinds have an important influence on technological developments and on the decisions of policy makers at the macrolevel. Another reason why these planning problems are so difficult is that political conflicts between spatial units or between policy objectives are intense. Geographical information systems have been designed to contribute to the solution of such planning problems. For this purpose, GIS systems have been supplied with various facilities for analysis, modelling and forecasting. Here, we consider the integration of GIS with another facility: multicriteria analysis (MCA), which aims at analysing the intensity and nature of conflicts between policy criteria, generating compromise alternatives and rankings of alternatives according to their degree of attractiveness. Thus, linking GIS with MCA enables policy conflicts to be analysed in a spatial context.

A short review of MCA is given in Section 12.2. In Section 12.3, a case study on agricultural landuse is introduced. Evaluation criteria are defined in Section 12.4. A multicriteria approach is applied in Section 12.5 to rank a set of alternatives and in Section 12.6 a linear programming approach is used to determine the optimal land use combination. Section 12.7 offers some concluding remarks.

12.2 Multicriteria analysis

During the early 1970s, a new class of evaluation analyses came to the fore, based on the fact that intangible effects and policy conflicts were seen to make up a central part of policy analysis (Cochrane and Zeleny 1973). This new class of methods has seen usually called multicriteria analysis (MCA). Multicriteria analysis can be conceived of as an extension of conventional evaluation methods. For example, cost-benefit analysis, with its reliance on market prices is now replaced by a much broader set of discrete MCA methods where prices (weights) may also reflect political priorities (Nijkamp, Rietveld and Voogd 1989). Linear programming, as another example, only deals with one objective function. In the context of MCA it is replaced by programming methods for multiple, conflicting objective functions (see Steuer 1986). This shift from one criterion to multiple criteria also means that the original concept of optimality is no longer the dominant concept. It is replaced by the concept of Pareto-optimality (or efficiency), leading to much broader sets of alternatives which are in principle acceptable. Reviews of the MCA literature can be found in Voogd (1983) and Nijkamp and Rietveld (1986). This section does not offer a comprehensive survey, but introduces those concepts that will be used in the remainder of this chapter. The discussion is restricted to discrete MCA problems, i.e. problems with a finite number of alternatives. The term alternative refers to any option available in a choice set. In a spatial context, an alternative may refer to a region as allocation for a certain activity. In notational terms, n can

H. J. Scholten and J. C. H. Stillwell (eds.),
Geographical Information Systems for Urban and Regional Planning, 129–139.
© 1990 *Kluwer Academic Publishers. Printed in the Netherlands.*

be used to indicate an alternative and j to indicate a particular criterion. Thus, p_{jn} can be interpreted as the effect of alternative n according to criterion j, where $n = 1,...N$ and $j = 1,...J$. The effects of all alternatives according to all criteria can be summarized by the evaluation matrix, P:

$$P = \begin{matrix} p_{11} & \cdots & p_{1N} \\ \cdot & & \cdot \\ \cdot & & \cdot \\ \cdot & & \cdot \\ \cdot & & \cdot \\ \cdot & & \cdot \\ p_{J1} & \cdots & p_{JN} \end{matrix}$$

Thus, the nth column of P contains the performance of the nth alternative according to criteria 1 to J. It must be noted, that it is not always possible to find quantitative values for the p_{jn}'s. Sometimes, only qualitative (nominal or ordinal) information on the effects of alternatives is available. One possibility of studying the degree of conflict between two criteria is by computing the coefficient of correlation between the rows concerned. A coefficient of correlation near to 1 would mean that there is little conflict between the two criteria: achieving a high value for one criterion does not prevent one from achieving a high value for the other criterion. The reverse would be the case with a correlation coefficient near to -1. Other ways of measuring the degree of conflict between criteria are discussed in Rietveld (1980). The evaluation matrix P can also be used to determine the so called ideal point. The ideal point is a vector combining the most attractive value for each criterion $j=1, ..., J$ as it can be found in each row of the matrix. The ideal point can be used as a point of reference to determine the attractiveness of alternatives, since those which are near to the ideal point are more attractive than other alternatives.

A weight vector, w, is another ingredient of many MCA methods. The weights $w_1, ..., w_J$ reflect the importance attached to criteria, as expressed by a decision making unit. A survey of methods to determine such a weight vector is given in Nijkamp and Rietveld (1986). It is clear that in many real world planning problems, there is considerable uncertainty about the quantitative values of the weights. One will often arrive at intervals, or rankings of weights. This may lead to a considerable uncertainty about the question of which alternative is the most attractive one. Various approaches exist to arrive at a final ranking of alternatives on the basis of P and w. The simplest approach is to use a linear utility function (weighted summation) for this purpose. Also, non-linear utility functions may be used (such as the multi-attribute utility theory of Keeney and Raiffa (1976). Another approach outside the realm of utility theory is concordance analysis (Roy 1968, Crama and Hansen 1983). Concordance analysis is a pairwise comparison approach entailing a detailed investigation of the pros and cons of alternatives for all pairs of alternatives. For reasons of convenience of presentation in this paper, only a simple weighted summation method is applied. Clearly, sensitivity analyses have to be carried out to check for the uncertainties about the weight vector, w.

12.3 Case study: agricultural land use

The agricultural sector in the Netherlands has experienced turbulent developments in the second half of the 1980s. Agricultural pricing policies of the European Community have led to chronic excess supply for certain product groups. A change in policy was inevitable, leading to a sudden cessation in the growth of output, or as occurred in the dairy sector, a decrease of output. Another problem has been the steady growth of manure production in animal husbandry. Leading to environmental problems in certain regions and inducing policy measures aimed at reducing manure production.

Future developments are uncertain: much depends on the agricultural policy of the European Community. Nevertheless, a real possibility exists that the agricultural sector will need considerably less land in the future. Technological development has been very fast in the past, leading to a rapid increase in output per unit of land. The opportunities for a further increase are not yet exhausted. However, the demand side is not very promising for most groups of agricultural products. The combination of stagnating demand for important product groups, and of a further increase in output per hectare would lead to an excess of agricultural land. This excess of agricultural land may be much larger than the amount of land to be used for standard non-agricultural purposes such as city growth or road construction (Douw et al. 1987). Several responses are possible in the agricultural sector, such as the introduction of new products, reorientation of technological development into a more land extensive approach, and the production of agricultural inputs which were formerly imported. Nevertheless, one cannot rule out the possibility that for certain agricultural areas of substantial size an alternative type of land use has to be found. Two types of alternatives seem to be relevant. One is a change of land use towards forestry or the development of natural areas. The other is a combination of the present agricultural use with a number of other activities leading to a less intensive use of the areas. Among these are labour intensive activities of farmers to increase the quality of the landscape or the ecological diversity. This would lead to a certain decrease in agricultural output. Farmers would receive a compensation for their activities from the government. Thus the policy questions arising here relate to:

(i)   the choice of strategy: change of land use versus combined land use, and
(ii) the choice of regions which are most suitable for these strategies.

In this policy problem two types of conflicts are dominant: conflicts between regions, and conflicts between agricultural interests on the one hand and interests in the field of recreation, natural environment, etc., on the other.

12.4 Definition of criteria

The present study will be carried out at the level of 118 agricultural regions in the Netherlands. The data used are retrieved from a GIS in use at the National Physical Planning Agency of the Netherlands (cf. Padding 1987). A complete account of the results of this study can be found in Van Herwijnen, Janssen and Rietveld (1989). Many factors play a role in the suitability of an area for a change or a combination of landuses. Two main classes of factors can be distinguished: 'opportunities' and 'demands'. Opportunities refer to the present use of the agricultural land; they represent the interest of the land to the agricultural sector. For example, opportunities to change the use of agricultural land are low in an area with a highly productive agricultural sector or with high prices of agricultural land. Demands refer to the potential of the agricultural land for other kinds of use (or a combination of uses). For example, demands depend on the need for recreational areas in the region and the suitability of the land for forestry development or the formation of national landscape parks.

In order to determine opportunities and demands, an evaluation matrix, P, has been formed consisting of 118 columns (regions) and 18 rows (criteria). Of the 18 criteria, 9 are related to the opportunities side, and 9 to the demands side. For both variables, (and both actions, i.e. change and combination) a summary indicator is constructed based on the weighted summation approach discussed previously. The weights have been determined on the basis of the views of a committee of experts in this field. In order to overcome the problem that the criteria are measured on different scales, the original data have been transformed in such a way that the score is 100 for the most attractive region and 0 for the least attractive region. This is only one of the possible ways of standardizing criteria (see also Rietveld, 1980).

The weighted summation procedure leads to a considerable reduction of the size of the evaluation matrix, P. Instead of the original 18 rows, only 4 rows remain which represent opportunities and demands for a change of landuse, and opportunities and demands for combined landuse. Figure 12.1 gives a graphical illustration of the results of the weighted summation procedure for the first 40 regions. Both demands and opportunities with respect to a combined land use look very good in Salland-Twente (region 34). Similarly, opportunities for a change in landuse are high in the Smilde region (28). The demands for a change in land use are low to very low in general agricultural areas including Humsterland (region 5).

Figure 12.1: Alternative land use utility scores for opportunities and demands, regions 1 to 40

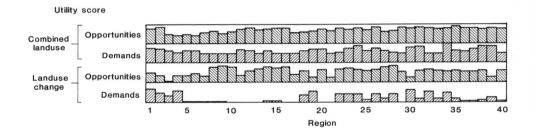

One must be aware that the results thus far depend on various assumptions about the weights, the method of standardization and the specific MCA method used. The MCA package used in this case, DEFINITE, (Janssen 1989) allows sensitivity tests to be carried out for this purpose. It must be noted, however, that the most important political question of how to weight agricultural interests compared with the natural environment or forestry interests, has not yet been dealt with. This problem of assigning weights to 'demands' versus 'opportunities' will be discussed in the next sections.

Our analysis has led to summary indicators for opportunities and demands based on a number of underlying criteria. In GIS systems, such a summary can in principle be made by means of overlays. Overlays have the advantage that they give a direct link between the underlying variables and the summary indicator but the overlay approach has its limitations. Firstly, overlays are difficult to use when there are many underlying variables (say, more than 4). Secondly, the overlay procedure does not enable one to take into account that the underlying variables are not equally important. Thirdly, the problem of defining threshold values will arise in evaluation studies. For example, to represent the variable agricultural income one has to fix a level x. This enables one to distinguish regions with a high income level (larger than x) from regions with a low income level (smaller than x). In such a case, the results of an analysis based on overlays may depend strongly on the choice of the threshold values. Fourthly, it is clear that the use of threshold values may lead to a substantial loss of information, since continuous variables are transformed into nominal ones. It follows, therefore, that the use of overlays for evaluation studies is in most circumstances not recommended. The use of MCA based summary indicators seems to be a better approach because:

(i)   they are easy to handle with many criteria;
(ii)   differences in importance between criteria are taken into account;
(iii)   there is no need to formulate threshold values; and
(iv)   there is no loss of information due to a down-scaling of continuous variables on a nominal scale.

12.5 Ranking 118 agricultural regions for combined land use and change of land use

In this section all 118 regions will be ranked according to their suitability for combined land use, and their suitability for change of land use. These rankings will be derived using multicriteria analysis.

Ranking of the regions for combined land use

The opportunities and demands for combined land use, as listed in the evaluation table (Figure 12.1), are shown again in Figure 12.2a. The demands scores represent the benefits that can be obtained from combined land use through nature development and recreation; the opportunities scores reflect the costs of combined land use for agriculture. The ideal region for this policy combines maximum demands (highest benefits) with maximum opportunities (lowest costs). This ideal region would have score (100,100) and would be found in the top right corner of the diagram. The regions are ranked on the basis of their distance to the nonexistent ideal region: the region closest to the ideal region ranks as number 1, the next closest as number 2 and so on. The total distance is measured as the sum of the horizontal and the vertical distance from the ideal region. Region 30 (Salland-Twente) ranks clearly first; the differences in distance for the next few regions are fairly small. Region 89 (the Westland), on the bottom left of the diagram, ranks by far as the most unsuitable region for combined landuse. The correlation coefficient shows to what extent opportunities and demands for combined land use coincide or conflict. A value close to 1 indicates minimal conflict between opportunities and demands; regions with high demands offer good opportunities. A value close to minus one shows extreme conflict: regions with high demands offer no opportunities and vice versa. The value 0.346 in indicates that opportunities and demands partially coincide. The ranking for combined land use is derived giving equal weight to agriculture and nature. The positive value of the correlation coefficient indicates that this ranking is fairly insensitive to changes in these weights. However, the ranking of the regions for change of land use is, as will be shown now, very sensitive to changes in these weights.

Ranking of the regions for change of land use

Opportunities and demands for change of land use in all 118 regions are shown in Figure 12.2b. The line in this diagram intersects the opportunities and demands axis at the same distance from the ideal region. This reflects that equal weight is given to nature and agriculture. All points on this line have the same position in the ranking. The regions can now be ranked visually by moving the line from top right to bottom left. The first to cross the

Figure 12.2: Opportunities and demands for combined land use and land use change

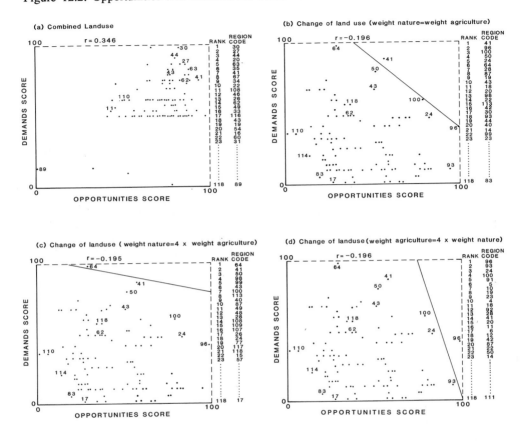

line and therefore the most suitable region is region 41 (the Oost Veluwe). It is clear that this region is closely followed by region 96 (West Zeeuws Vlaanderen), and region 100 (the Biesbosch). Note also that the points in this diagram are fairly evenly distributed. The negative value of the correlation coefficient (-0.196) indicates that the ranking will be sensitive to changes in weights. In Figure 12.2c the weight assigned to nature is four times the weight given to agriculture. The distance of the intersection with the opportunities axis is now four times the distance of the intersection with the demands axis. This implies that a region with high demands will rank high even if the opportunities are limited. Region 64 ('t Gein), a region with maximum demands but fairly low opportunities, is now the first to cross the line closely followed by region 41 (the Oost Veluwe). Quite a different ranking results if the weight given to agriculture is four times the weight of nature. The ranking as shown in Figure 12.2d is now headed by region 96 (West Zeeuws Vlaanderen), followed by three regions on almost the same position: Noord Beverland (93), Smilde (24) and Biesbosch (100).

Map representations

Each weight combination results in a list of 118 regions with the most suitable region at the top, the least suitable region at the bottom and all other regions, according to their suitability, in between. This offers precise information on the position of each region in the ranking but offers no insight in the spatial pattern of the ranking. Therefore the rankings can also be shown on maps. The loss of detailed information on these maps is compensated for by insight given into the relation between position on the map and position in the ranking. The influence of different weights on the spatial pattern of the ranking can be analyzed using the three maps presented in Figure 12.3. The number of regions in each group is equal on all three maps. The change of weights results in a shift of the 20 regions over the map while keeping the number in each group constant. A shift in the weight assigned to nature from 0.80 in (a) to 0.20 in (b) results in a shift of the position on the map of a large number out of the 20 highest ranked (black coloured) regions, in such a way that regions move from the centre (a) to the north of the country (b).

Note that the regions in the east of the country rank within the best 20 regions if equal weights are given to nature and agriculture and between 20 and 40 if the weight of nature is either increased or decreased. This indicates a balance between opportunities and demands in these regions. The representations can easily be adjusted to specific policy questions focusing on one region, one aspect, focus on the best or the worst regions, and adjustment of the level of aggregation (see Herwijnen et al. 1989, Tufte 1985).

12.6 The optimal land use combination: an integrated approach

In the previous section, the regions were ranked according to their suitability for either combined land use or change of land use. No tradeoff was made between these two options. In this section both options will be dealt with simultaneously . The area allocated to each type of landuse in each region is also calculated, and the solution is made subject to constraints such as the available budget for a policy option. The aim of the approach is to calculate the optimal feasible combination of landuses. This optimal solution can be found with the help of a linear programming approach (see Killen 1983, Steuer 1986).

The utility function

For each region a decision needs to be taken about the percentage of the region to be allocated to change of land use, combined land use, and agriculture. Since there are 118 regions this results in 118 x 3 = 354 decision variables. The utility of region i is defined as follows:

Size of agricultural land in region i x
  (Percentage x utility of combined land use in region i
    + Percentage x utility of change of land use in region i
    + Percentage x utility of agriculture in region i)

Priorities can be expressed as weights to the utility indices. The utilities per region for each type of land use are derived from the utility scores calculated in Section 12.4. Total utility is calculated as the sum of utilities of all regions. The aim of the analysis is to calculate values for the decision variables (the percentages) in such a way that the overall result or total utility is as high as possible.

Figure: 12.3: Map representation of ranking land use change under different weighting
assumptions

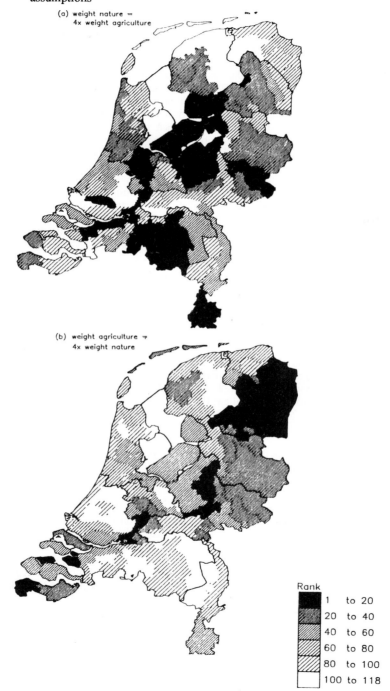

(a) weight nature =
    4x weight agriculture

(b) weight agriculture =
    4x weight nature

Rank
1   to 20
20   to 40
40   to 60
60   to 80
80   to 100
100 to 118

Constraints

Optimization of the utility function is subject to constraints per region and constraints on totals for all regions such as total budget available. The second type of constraints result in the problem of the optimal allocation of policy effort over all regions. This type of constraints creates the necessity to use a linear programming approach. To prevent unrealistic levels of change of land use in each region a maximum of 40% of a region can be allocated to combined land use, and a maximum of 20% of a region can be allocated to change of land use. The sum of agricultural, combined and change of land use should of course be 100%. An overall maximum of 100,000 ha is set for the total area of combined land use as well as a maximum of 100,000 ha for change of land use. At least 100,000 ha of agricultural land must be reallocated. Finally, the total budget available for purchases of land is set to DFL 10 billion.

Optimization

A linear programming routine is used to find the percentage combination of the three types of land use in all regions that maximises total utility and meets all the constraints. The percentages for combined land use and change of land use are calculated first and the percentage for agriculture is calculated by subtracting these percentages from 100. In the Oostelijk Weidegebied, for example, 0% is allocated to change of land use, 38% to combined land use and 62% to agriculture. In the optimal solution the constraints set to the maximum area of change of land use and combined land use are both exactly met. This implies that 200,000 ha of agricultural land has been reallocated. Total budget spent equals DFL 5 billion which means that the available budget is not a limiting factor in this solution. A different solution is reached if the constraint on the maximum area for combined land use is relaxed from 100,000 ha to 200,000 ha. The total area for combined land use in this solution also equals the level of the constraint. The increase in combined land use results in an equal decrease of agriculture.

Figure 12.4: Sensitivity of land use to priorities (weight nature = 1 - weight agriculture)

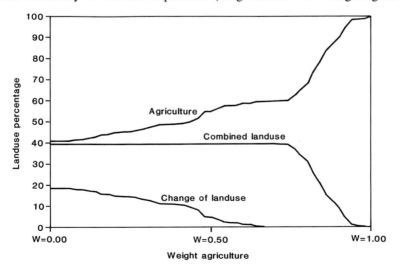

Weights

To reach these solutions, equal weights are assigned to nature and agriculture. The influence of these weights on the percentages allocated to the three types of land use is shown in Figure 12.4. The combined land use percentage remains constant as long as the weight of agriculture is between 0 and 0.74. This percentage sharply declines in favour of agriculture if the weight assigned to agriculture exceeds 0.74.

12.7 Concluding remarks

A wide variety of multicriteria methods are available to be applied in combination with GIS. The usefulness of this combination is illustrated by an application of a multicriteria approach to the reallocation of agricultural land in the Netherlands. In this application data available from the National Physical Planning Agency of the Netherlands (RPD) are successfully used to support the policy questions of where, and to what extent, the use of agricultural land should be changed or combined with other activities. Data available from a GIS are by their nature useful to support decisions with a spatial dimension. Multicriteria methods in combination with adequate map presentations of the results allow for an optimal use of these data. The present application is a first attempt. In the future more attention should be devoted to the construction of the indices, the aggregation level, and to the quality and retrieval of data from the GIS. The procedure described is available as a decision support system. This allows people involved in decision making to change the definition of the indices, the weights and the constraints according to their own opinions (see Herwijnen and Janssen 1988).

References

Cochrane, J.L. and Zeleny, M. (eds) (1973) Multiple Criteria Decision Making, University of South Carolina Press, Columbia

Crama, Y. and Hansen, P. (1983) An introduction to the ELECTRE research programme, In P. Hansen (ed) Essays and Surveys on Multiple Criteria Decision Making, Springer, Berlin, pp. 31-42

Douw, L., van der Giessen, L.B., and Post, J.H. (1987) De Nederlandse landbouw na 2000; Een verkenning, Mededeling 379, Landbouw Economisch Instituut, Den Haag

Herwijnen, M. van and Janssen, R. (1988) DEFINITE, A support system for decisions on a finite set of alternatives, Proceedings of the VIIIth International Conference on MCDM, Manchester

Herwijnen, M. van, Janssen, R. and Rietveld, P. (1989) Een multicriteria analyse van alternatieve aanwendingen van landbouwgrond, Institute for Environmental Studies, Amsterdam

Janssen, R. and Rietveld, P. (1985) Multicriteria evaluation of land-reallotment plans: a case study, Environment and Planning A, 17, 1653-1668

Janssen, R. (1989) Beslissings Ondersteunend Systeem voor Discrete Alternatieven, Beschrijving en handleiding, IVM, Amsterdam

Keeney, R.L., and Raiffa, H. (1976) Decisions with Multiple Objectives, Wiley, New York

Killen, J.E. (1983) Mathematical Programming for Geographers and Planners, Croom Helm, London

Nijkamp, P. and Rietveld, P. (1986) Multiple objective decision analysis in regional economics. In P. Nijkamp (ed) Handbook of Regional Economics, North Holland Publishing Co., Amsterdam

Nijkamp, P., Rietveld, P. and Voogd, H. (1989) Multicriteria Evaluation in Physical Planning, North Holland, Amsterdam

Padding, P. (1987) De toekomst van de landbouw in ruimtelijk perspectief, een toepassing van het GIS ARCINFO, Geografisch Instituut, Rijksuniversiteit Utrecht and Rijksplanologische Dienst, Den Haag

Rietveld, P. (1980) Multiple Objective Decision Methods in Regional Planning, North Holland, Amsterdam

Roy, B. (1968) Classement et choix en présence de points de vue multiples, RIRO, 2, 57-75

Steuer, R.E. (1986) Multiple Criteria Optimization: Theory, Computation and Application, Wiley, New York

Tufte, E.R. (1985) The Visual Display of Quantitative Information, Graphics Press, Cheshire, Connecticut

Voogd, H. (1983) Multicriteria Evaluation for Urban and Regional Planning, Pion, London

Ron Janssen
Vrije Universiteit Amsterdam
Instituut voor Milieuvraagstukken
Postbus 7161
1007 MC Amsterdam
The Netherlands

Piet Rietveld
Vrije Universiteit Amsterdam
Economische Faculteit
Postbus 7161
1007 MC Amsterdam
The Netherlands

PART V

SPATIAL ANALYSIS, MODELLING AND DECISION SUPPORT

13 THE APPLICATION OF GEOGRAPHICAL INFORMATION SYSTEMS IN THE SPATIAL ANALYSIS OF CRIME

Dean R. Anderson

13.1 Introduction

Urban crime is a major problem facing cities in the United States. Police and public safety departments continue to feel the stress of manpower shortages that must be continuously balanced against the rising crime rate. In addition to standard crimes such as burglary and theft, many departments have special groups to handle gang and drug related crimes. These groups are typically assigned to problem areas at specific times in response to problems. The tools available in a Geographical Information System (GIS) provide a method for crime analysts to examine crimes in a different way than with the traditional record keeping systems. The GIS tools are available on a variety of platforms including micro-computers, workstations, mini-computers, and mainframe computers. However, in order for the tools to be useful, they must fit into existing operations within police departments, interface with existing systems, and provide an easy way for crime analysts to perform spatial analysis. This chapter examines how a crime analysis application system can be developed, explains how the GIS tools are used, and illustrates the results of the development process.

13.2 Design methodology

System development requires that a structured methodology be followed. Many methodologies exist and in general all should:

(i) identify the user requirements including the activities performed and the data required to complete the activities;
(ii) design a conceptual and then physical system including data standards and user interfaces;
(iii) test and modify the designs using several prototypes;
(iv) incrementally code the system including development of documentation and full testing;
(v) install the system; and finally
(vi) review the system and fix any bugs or make minor adjustments.

Identifying user requirements involves understanding who each user is and how they can potentially use the system. A series of interviews or work sessions are held in which the following information is gathered:

(i) project goals and objectives;
(ii) users, tasks and organization descriptions;
(iii) all data sources;
(iv) inventory of examples of similar project outputs and data;
(v) frequency, quantity, and accuracy requirements of data; and
(vi) products needed to complete the project.

The design is used to construct a conceptual and physical view of the data, and determine the steps necessary to meet the study's goals and objectives. The design addresses each user

H. J. Scholten and J. C. H. Stillwell (eds.),
Geographical Information Systems for Urban and Regional Planning, 143–151.
© 1990 *Kluwer Academic Publishers. Printed in the Netherlands.*

based on information gathered in the user requirements step. The following information is gathered during the design phase:

(i)   data category definitions;
(ii)   pictures or sample maps of data categories;
(iii)   diagrams which illustrate inter-relationships of data;
(iv)   task descriptions for completing the analysis of crimes including conversion, creation, and display of crime information;
(v)   diagrams that illustrate inter-relationships of data categories and tasks;
(vi)   identification of user interfaces; and
(vii)   development of data structures.

Based on the design, a prototype system for a particular city is developed using available criminal incident data and the street network. The prototype system will test the user interface, the data flow from existing systems, and the potential hardware configurations. The prototype is presented to the police department staff for review. Based on the prototype the police finalise the hardware for operation, the user interface, and the display layouts for the system. Using the prototype and resulting comments from the police, the system is incrementally coded, tested, and documented. Coding is performed for each interface as identified in the prototype system. Incremental testing of the system is performed by the police department, so that they will have an active involvement.

When all functional groups for the system are completed, the system is installed as a complete application. The police department is given training, user and programmer documentation. Training addresses both user applications and system maintenance. Any initial problems found with the system are fixed or discussed. Based on an initial review of the total system, minor modifications are made. The system now becomes the property of the police department and they are responsible for all future maintenance and enhancements.

13.3 Available tools

Three common GIS analysis tools that may be applied to the spatial analysis of crime are overlay, address matching and network analysis. This section briefly discusses each tool.

Overlay tools allow users to create a composite of geographic features from independent geographic sources that occur in the same geographic area. These overlays use standard graphics theory. Figure 13.1(a) illustrates how a basic overlay works. Another type of overlay is to aggregate points within polygons to create summaries of point information within a polygon or to identify within which polygon a point lies. This function allows users to create summaries of criminal incidents by management district or to identify in which management district a crime is located. Figure 13.1(b) illustrates this concept.

Address matching or geocoding allows users to create geographic features by matching addresses from a tabular file to a relative position on a street segment that contains an address range. This function will allow users to create point features from addresses such as crime locations, student locations, or permit applications. The concept is illustrated in Figure 13.1(c).

Two techniques are often available for network analysis (Figure 13.1(d)): routing and resource allocation. Routing determines the optimum paths for movement of resources through a linear network. Examples of a linear network are: roads, streams, gas lines, water mains, and electrical circuits. Locating an optimal path through a network must be based on impedance to flow along a stretch of road, such as a speed limit, as well as at intersections. An example of a typical routing problem is to find alternative paths for rescue vehicles in a city during peak traffic hours. Resource allocation finds the nearest centre for each segment through a linear network. The resource allocation is based on impedance to flow through a linear network as route does and will also allow resources to accumulate at resource centres. An example of a typical allocation problem is to locate all street segments that may be reached by a fire truck within five minutes.

Figure 13.1: Common GIS analysis tools

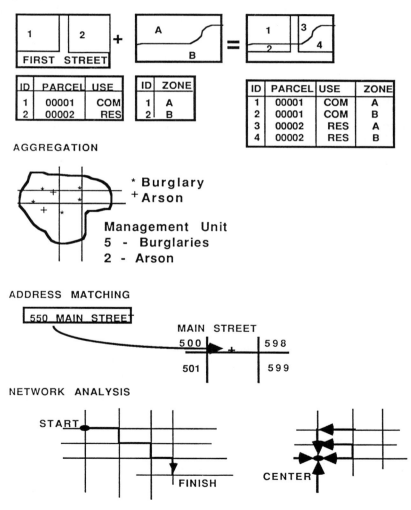

13.4 Results

The application developed is called the Crime Analysis Mapping System (CAMS) and is maintained by the police department. Potential users of the system are:

(i)   detectives and field officers;
(ii)   the crime analyst;
(iii) research and development staff;
(iv)   city management; and
(v)   public support groups.

These users have a wide variety of needs. Examples of their requirements are as follows:

(i)   Each week the existing reports system is used to produce a summary report of incidents. A map displaying the associated incident locations is required.
(ii)   Beat and general management boundaries are difficult to maintain and often require the department to hire expensive outside consultants. This activity should be completed by the department.
(iii) The department should continually examine the distribution of manpower and special personnel relative to crime activity.
(iv)   The department should be able to examine criminal activity near specific businesses such as all-night convenience stores, liquor stores, or gas stations.
(v)   The department must provide crime summaries in response to public inquiry about general crime activity in neighborhoods or crime-watch zones.
(vi)   Field people can obtain reports about existing crime activity. Maps that address individual crimes are also required.

A major task required in developing the system is the identification of what information is going to be in the system. Five data entities are identified as follows:

CRIMES: A single criminal incident
PEOPLE: Any person involved in the crime
LOCATION: Address location where the crime occurred
MANAGEMENT DISTRICT: A geographic area for manpower management
PROPERTY: Any property affected by the incident

The information about the crime incident is often stored in dispatch and record keeping systems. Sometimes these two systems are independent and at other times are closely linked. Dispatch systems typically have good information about the response time for an incident and are kept up to date but have sketchy information about the crime. Record keeping systems have good information about the criminal incidents but are usually one to several days old.

Many cities also keep information about major street arterials, streets with addresses, and management boundaries within existing geographic information systems. This information could be used to support the criminal analysis functions. For example, a pin map of crime locations is more meaningful if it is produced with streets and street names. By using existing data within the city, the police can make substantial cost savings.

The data for the two systems is ported as flat ASCII files from the record keeping and dispatch systems. Once converted, each incident is address matched to a point location along the street network. Approximately 95% of the incidents can be located this way. The remaining five percent must be matched manually. The point locations must then be overlaid onto the management boundaries and summary statistics developed. The general flow of this process is described in Figure 13.2. Once loaded into the application, the point locations are then viewed. Several user interfaces are possible for examining data and screen menus appear to meet the general user requirements. The general screen layout and an example of the menu are displayed in Figure 13.3.

Figure 13.2: The process of generating a map of crime locations

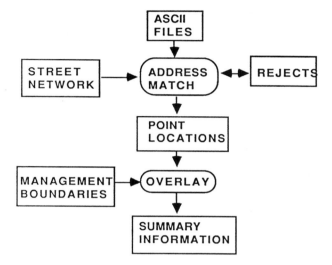

Figure 13.3: An example of the user interface

COMMANDS	REGION	CRIME	ATTRIBUTES	DRAW	SUMMARY	REPORTS	SYMBOLS

TACOMA   POLICE   DEPARTMENT

Clear - Map
    - Legend

Enter Arcplot   Pull Down Menu

Quit

Reset Page

Area For Maps

LEGEND

Legend
or Map Key

SYMBOLS

+   /\/

Locator
Map

DATE: 6/10/86

Figure 13.4: User documentation for the crime analysis system

CRIME ANALYSIS SYSTEM

DRAW   MENU

This option will allow you to draw selected features.

COMMANDS	REGION	CRIME	ATTRIBUTES	DRAW	SUMMARY	REPORTS	SYMBOLS

Draw points for all selected crimes using the current marker symbol.

Draw points for all selected crimes using predefined symbols.

Draw all streets within the selected region.
Draw all arterials.
Draw all arterials and arterial street names.
Draw all district and subdistrict boundaries.
Draw all district and subdistrict boundaries and nos.
Draw all census block boundaries.
Draw all census block boundaries and numbers.

Draw the north compass arrow.
Redraw the map or screen borders.
Use a second menu to use some general drawing tools.

Cancel this menu.
Obtain help for this menu.

DRAW

CRIMES

CRIMES W/SYMBOLS

STREETS
ARTERIALS
ARTERIALS W/NAMES
DISTRICTS
DISTRICTS W/NBRS
CENSUS BLOCKS
CENSUS BLKS W/NBRS

COMPASS NORTH
PLOTTER BORDERS
MISC DRAWING TOOLS

CANCEL
HELP

SCREEN COMMANDS

TEXT - SMALL
TEXT - MEDIUM
TEXT - LARGE

LINE
BOX
CIRCLE
MARKER

LISTING FULLSCREEN
THREE LINES

CANCEL
HELP

Place a small size text string on the graphic screen.
Place a medium size text string on the graphic screen.
Place a large size text string on the graphic screen.

Draw a line using the cursor. (to end the line type <CNTL> and the last button on the right of the cursor.)
Draw a box using the mouse by entering its min and max.
Draw a circle by entering its center and a point on the perimeter.
Draw a marker on the graphic screen using the cursor.

Change format of reprorts or listings to be full screen.
Change format of reports or listing to be three lines.

Cancel this menu.
Obtain help for this menu.

The user interface for the application is constructed as a set of high level tools that allow users to custom build reports and maps. It is based around the following:

(i)   select an area for mapping;
(ii)   select and map a subset of criminal incidents based on major criminal incident categories such as theft, burglary, or traffic accidents;
(iii) select and map a subset of criminal incidents based on any attributes such as response time, time of incident, method of operation or premise:
(iv)   draw all incidents using predetermined symbols or user assigned symbols for streets, arterials, management boundaries, and incidents;
(v)   produce shaded maps of crimes by census block or management unit;
(vi)   list and obtain formatted reports for all attributes; and
(vii) change mapping symbology.

User and program documentation are important aspects of the system. Documentation includes program calls and examples of menus as shown in Figure 13.4.

The proof of any application system is whether it can answer questions in a meaningful way without taking a lot of time. The following examples illustrate four types of output generated in response to typical questions. The examples use fictitious datasets used in prototype development for the City of Tacoma, Washington. Figure 13.5 indicates a shaded map of the total number of thefts by census block in the Second Precinct; Figure 13.6 shows a map of all crimes that occurred in a single census block (block 113); Figure 13.7 indicates a map of all crimes that occurred within 1/2 mile of a suspect's house with an address of 6020 N 45th ST; and Figure 13.8 presents a dot map of thefts that occurred between midnight and six o'clock. All service station crimes are highlighted as stars and the method of operation for each crime is listed.

Figure 13.5: Example 1: Total thefts by census block

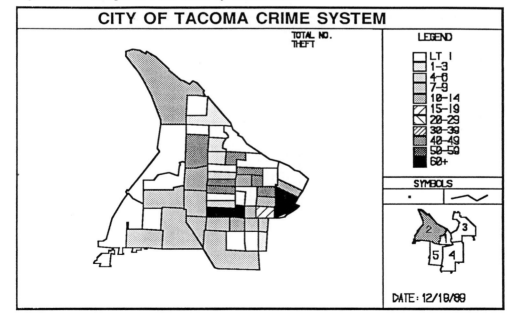

Figure 13.6: Example 2: Crimes within a single census block

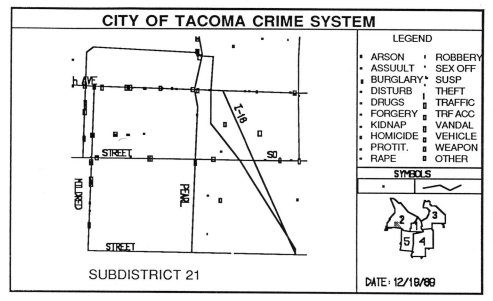

Figure 13.7: Example 3: Crimes within specified distance of one location

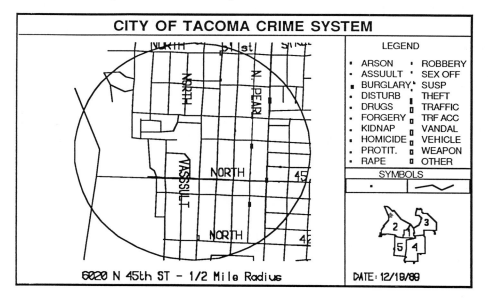

Figure 13.8: Example 4: Thefts from service stations

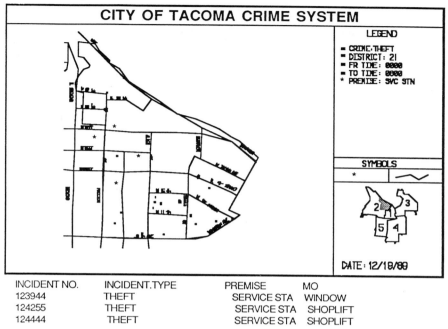

INCIDENT NO.	INCIDENT.TYPE	PREMISE	MO
123944	THEFT	SERVICE STA	WINDOW
124255	THEFT	SERVICE STA	SHOPLIFT
124444	THEFT	SERVICE STA	SHOPLIFT

13.5 Conclusion

The tools available with a geographic information system can add new functionality to traditional crime record keeping systems. Maps, geographic queries, and related reports add a new dimension to the analysis of criminal information. These tools must be incorporated into application systems that fit into the existing department. The analysis system, like any other system, must be developed with a good design methodology. It must also have an easy-to-use interface, good documentation, and must work. It is currently being used by several cities in the United States including the City of Seattle and the City of Tacoma. The tools used to perform criminal analysis can also be applied to a wide variety of other urban problems such as emergency response, permit applications, fire incident analysis, and zoning. The applications can be as varied as the user's imagination.

Dean R. Anderson
Environmental Systems Research Institute (ESRI)
525 North Columbia Street Suite 205
Olympia
Washington 98501
USA

14 SPATIAL ANALYSIS AND GEOGRAPHICAL INFORMATION SYSTEMS: A REVIEW OF PROGRESS AND POSSIBILITIES

Stan Openshaw

14.1 Introduction

The rapid expansion of GIS and the increasing availability of digital map data is resulting in a geographic data explosion of quite unparallelled historic proportions (Rhind et al. 1989). GIS has an exceptionally broad base of interested users covering a wide spectrum of disciplines and interests and has been successful because it meets basic common needs for geographic information technology. However, this geo-processing industry is deficient in the almost total absence of spatial analysis functionality. It is true that the broadly based cartographic origins to GIS placed far more emphasis on map manipulation than on map analysis and that there has been a failure to go beyond the map when contemplating analysis. Nevertheless, it is only a matter of time before attention will move on from the necessary but boring digital map database creation era with typically focused but limited management information uses, to one in which more emphasis is placed upon more general corporate uses of geographic information. It is inescapable that modelling and other spatial analytical uses of the data will gradually become more in demand. However, it is not a one-way process and there is need both for new analysis functions to be added to the GIS toolbox and also new data structures to support their development and application.

There is a danger that the growing imbalance between the availability of geographic data and the limited range of analytical technology will slow the growth of GIS and result in a failure to make full use of the information being collected. This problem exists at a time when the increasing convergence between cartographic and bureaucratic computer systems, made possible by GIS, is also creating many new opportunities for spatial analysis. In many instances, the purpose behind such analysis is vague and relates mainly to the fact that the data exist rather than to a more traditional form of highly focused scientific inquiry. The emerging mountain of real and potentially creatable geographically referenceable data challenges the conventional manner by which statistical analysis and modelling have been performed in the past, particularly in a geographical context. The challenge can be viewed as involving the need for an automated and more exploratory modus operandi in a situation which is data rich but increasingly (by comparison) theory poor. Put another way, the historic emphasis on deductive approaches is becoming less practicable because of the increasing dominance of data-led rather than theory-driven questions. There is a need for a more creative approach and for analytical tools that can suggest new theories and generally support data exploratory functions. Whether these new methods can only be based on a rigid application of traditional scientific experimental designs is a matter for subsequent debate. Suffice it to say that maybe GIS only needs simple technology. The limitations on spatial analysis combine with the usual absence of process detail to argue strongly in favour of a more relaxed, flexible, artistic and less statistical approach than perhaps many may have expected.

There are many practical problems. Most of geographical data is inherently difficult to handle; for instance, data for points and irregularly shaped and defined zones are far more difficult to analyse in a mathematically rigorous fashion than, for example, lattice data from a remote sensing device when various simplifying assumptions can be made. Geographical data is also

153

H. J. Scholten and J. C. H. Stillwell (eds.),
Geographical Information Systems for Urban and Regional Planning, 153–163.

very seldom 'clean', in that fuzzyness often affects both resolution and representational accuracy in a difficult to handle manner. Possible analogies with sampling errors are not appropriate or applicable. Furthermore, many GIS operations seem to actually create spatial patterns (by aggregation with the zoning system acting as some kind of filter and pattern detector). Other GIS procedures (viz overlays) can add additional levels of error and also propagate errors thereby contaminating previous 'clean' data sets. There is probably not much that can be done to eliminate all sources and causes of geographical error and uncertainty in the short term, or indeed, ever. Currently, their real magnitude and impact is relatively little understood. What is more obvious, is the need to consider how methods of analysis can be devised that might be able to handle these problems, rather than simply ignore them so that existing techniques can be continued to be applied.

It is these and other endemic problems that characterise most GIS relevant data, that has seemingly created a widening gap between what 'spatial statistics' as viewed from a statistical perspective can offer and 'spatial statistics' as viewed from a GIS perspective needs to be able to offer to meet current demands from these systems. The traditional spatial statistical focus on forests, plants, sea anemones, drumlins, birds, etc, rather than people-based data, and perhaps also on military rather than civilian problems, tends to reflect their greater tractability to mathematical analysis. The problems of analysis in a GIS context are both extreme and very important. If statisticians are to remain in touch in this area then they need to start discovering how to handle 'dirty data', to develop GIS-relevant geographical analysis technology, and at least to start to attempt to answer the questions that typical users of the data are prone to ask. Likewise, if geographers want to do any better then they should:

(i)   stop trying to be second rate statisticians;
(ii)   throw away most of the spatial analytical, technology inherited from the 1960s Berry and Marble (1968) era; and
(iii)   rethink precisely what spatial analytic techniques and GIS needs, identify the most appropriate ones, and then put in the development effort necessary to make them appear sooner rather than later.

The purpose of this review is to argue the need for developing more relevant and appropriate spatial analysis technology for inclusion in tomorrow's GIS. It is time to 'burst free' from the constraints of the past and develop from first principles the kinds of geographical analysis technology needed to make best use of the tremendous opportunities being created by GIS developments.

14.2 Whither spatial analysis?

It is important not to neglect the lessons from the first spatial analysis driven quantitative revolution in geography. Spatial analysis technology is extremely limited in what it can offer. Many people find that pattern analysis and description is not particularly useful in their search for process knowledge and causal understanding. Such people, and there are many of them, probably suffer from unrealistic expectations of what geographical analysis can provide and what quantitative social science research can actually deliver. It is important also to understand that spatial analysis is not an end in itself but a tool that is only useful in the context of some kind of application. As such spatial analysis is problem driven and is really part of what is often called spatial decision support techniques. It follows, therefore, that spatial analysis is merely another class of map data processing technology which, if generic functions can be defined, has the potential to further enrich the GIS ensemble.

The problem is where to start. It is apparent that the developers of GIS saw little need to put much effort into spatial analysis technology, with only a few exceptions (e.g. SPANS from TYDAC). Maybe this is a reflection of their AM/FM or LIS or cartographic origins. Whatever the reason it is now a major omission. One obvious reaction is to simply import the available spatial analysis technology as standard functions inside GIS systems. There are, however, a number of problems with this approach:

(i)   a lack of standard methods expressed as portable software;
(ii)   a lack of GIS-adequate spatial analysis technology;
(iii)   the need to develop analysis tools that can handle error in spatial databases; and
(iv)   no clear view of what spatial analytic functions are needed within GIS.

The fourth issue really needs to be urgently addressed. It is obviously impossible to embed the whole of spatial statistical technology (for what it is worth) plus all available statistical methods into a GIS; even if some researchers are actually trying to interface with complete statistical packages (e.g. GLIM)! Instead, a far more restricted approach would appear more reasonable. Attention needs to be focused on defining a small set of generic spatial analysis functions that can be built-in as standard GIS operations and, also, define another set of analysis tools that would seek to provide new analytical functions which are only possible within a GIS environment. The latter class of methods may well have a GIS included within themselves.

These varying perspectives are shown in Figures 14.1 to 14.3. Figure 14.1 shows the missing link hypothesis. It is argued here that embedding statistical packages (e.g. GLIM or SAS) in a GIS or vice versa is not an adequate solution because the need for spatial analysis functionality cannot be simply met by importing 'willy nilly' complete statistical packages. This claim reflects the view that there is something special about the nature of geographic data and the needs of geographic analysis which are not met by conventional statistical analysis. Accordingly it might even be argued that the absence of any easy bridge as shown in Figure 14.1 at least has the property of discouraging the use of inappropriate technology. Figure 14.2 articulates the built-in spatial analysis function approach. The problem revolves around deciding which spatial analysis functions and operations are sufficiently general and generic to justify the effort. It also assumes a particularly simple relationship between GIS database and analysis methods (e.g. no extensive or special spatial data processing needs) which is probably unrealistic.

Figure 14.1: Missing link hypothesis

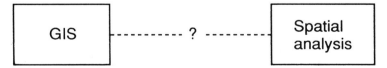

Figure 14.2: Built-in spatial analysis functions

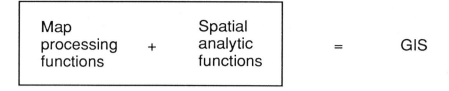

Figure 14.3 illustrates another, complementary, alternative in which the GIS, or rather some aspects of it, are an integral part of a spatial analysis technique. The result is a standalone, special function, system. It is unlikely to be able to use an existing GIS toolbox because the main justification for this separatist approach is to meet the need for extremely rapid spatial data processing and, possibly to utilise the power of a modern supercomputer. This strategy may also appear somewhat divisive but distributed computing and the growing power of workstations should allow the seams to become invisible and provide some hope that physical integration is not far away. Meanwhile, it would be silly to neglect the opportunities for developing new spatial analysis methods using supercomputers and parallel processors mainly on the spurious grounds that just now there is no GIS that will run on the hardware.

Figure 14.3: Building a spatial analysis method that includes a GIS

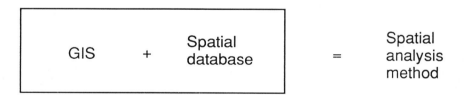

14.3 Criteria for GIS relevant spatial analysis methods

In addition to the problems of knowing which spatial analysis methods should be incorporated into GIS from the existing set of tools, there is also the practical necessity to make sure that the methods can actually work with, or in, a GIS environment. Criteria can be defined which provide basic design objectives for all developers of new spatial analysis methods and for supporters of old methods who wish to import their pet tools into GIS. This is useful to remove some of the inappropriate and inapplicable left-overs and rubbish from the geographical literature before they can be re-born or re-invented two generations later supposedly for GIS applications. Methods that never did work well enough in the 1960s will certainly not work any better in the 1980s, unless there were special computational or data dependent reasons for their failure. There is no great virtue in merely installing failed and otherwise inadequate spatial analysis methods from the 1960s into the GIS designed to be operational in the 1990s.

A number of GIS relevancy criteria can be defined as a basic checklist. This is important at a time when virtually everything and anything sitting on a computer that is even remotely related to geographical data is being re-labelled as being GIS relevant. In a real GIS environment, as distinct from a pseudo and fake one, the following features should be defined as important for any spatial analysis that is meant to be GIS useful:

(i)   methods should be able to handle large data sets (say 10,000 or more spatial objects) without any fundamental difficulty;

(ii)   methods should be explicitly geographical and make some use of the unique features that characterise spatial data as being different from non-spatial data;

(iii)   there should be no barriers to eventual portability of the technology;

(iv)   new methods should be potentially capable of being interfaced or tightly coupled into GIS systems or vice versa;

(v)   they should address issues and areas of perceived applied and/or vendor, importance rather than seek to inform abstract academic debate;

(vi)   there should be some focus on generic rather than idiosyncratic topics for which there is general interest and support;

(vii)   explicit attention should be given to the handling of positional errors and other kinds of spatial data uncertainty;

(viii) attention needs to be given to the explicit incorporation of temporal dynamics;

(ix)   the techniques need to be highly automated to make good use of powerful computational environments and reduce dependency on manual intelligence; and

(x)   deliberate attempts should be made to exploit the increasing computing power to develop computational (i.e. simulation) rather than analytical (i.e. assumption-ridden) spatial analysis methods.

14.4 Developing generic spatial analysis functions

It is also important to try and develop a new style of analytical GIS technology which can make good and appropriate use of the geo-data rich environments that GIS is providing in so many different areas of application. This task is itself an international problem with an extensive research agenda. The US National Center for Geographic Information and Analysis (NCGIA 1989) is clearly aware of the problem and the spatial analysis theme is implicit in several of their first three years worth of research initiatives; see Table 14.1.

In the UK, the ESRC's Regional Research Laboratory (RRL) initiative identified spatial analysis as one of three major research objectives for the eight RRLs to investigate. Attempts have also been made to identify which spatial analysis methods are receiving most attention (Table 14.2). From this broad list, five self-explanatory areas where key research is needed have been identified in Table 14.3 (Openshaw 1989a). This is clearly only one of a number of different spatial analysis research manifestos but it is often useful to speculate in this way as a means of stimulating additional research.

14.5 Exploratory geographical analysis

Inherent in these lists of areas where further research is needed is the recurrent theme of Exploratory Geographical Analysis (EGA). There is a growing requirement for analytical methods which are capable of exploring, summarising, and otherwise exploiting new geographic databases for a variety of purposes. It may seem that this task constitutes little more than the geographic description of map patterns rather than 'explanation'. This is quite true! Precise quantitative explanation of the patterns and processes found in geographic systems based on available geographic data is usually, if not always, impossible. There are definite limits on the usefulness of geographic information and on the value of a geographical style of investigation. Map displays and map analysis are useful but only during the initial stages of inquiry. The map is a wonderful communication device but it can also mislead, hence one role for spatial analysis is to process the data to minimise the potential problems of over-enthusiastic map users.

Consider an example from spatial cancer epidemiology. It has been suggested that there are five key questions that need to be answered about cancer patterns:

(i)   is there evidence of clustering?
(ii)   where are the clusters?
(iii)   how confident can we be that each identified cluster is real?
(iv)   how frequently might clusters occur in a particular subregion purely by chance? and
(v)   what background socioeconomic and environmental factors might be responsible?

Table 14.1: US NCGIA research initiatives

Initiative		
Year one	1*	Accuracy of spatial databases
	2	Language of spatial relations
	3	Multiple representations
	4	Use and value of geographic information in decision making
	5	Architectures of very large GIS databases
	6*	Spatial decision support systems
Year two	7*	Visualization of the quality of spatial information
	8	Expert systems for cartographic design
	9	Institutions sharing spatial information
	10*	Temporal relations in GIS
	11*	Space-time statistical models in GIS
	12*	Remote sensing and GIS

Source : NCGIA (1989)
Note: * topic related to spatial analysis

All five questions concern pattern description and related inference, there is no obvious means of explaining the observed patterns only some probability that patterns exist. Whatever is causing the patterns is a matter for subsequent study at a different scale, using different methods, by expert epidemiologists. It is not a matter for geographers or for GIS. The role of spatial analysis should be restricted to questions that the available geographic information might be capable of answering. Sometimes this role will constitute no more than the first stage of a more extensive study with the GIS role being limited to providing an indication as to where further research should be performed. This is still an extremely useful, if limited, role.

Hypotheses obtained from the analysis of data for one region can be tested (with considerable power) in another. Pattern description and inductive styles of analysis may be difficult but they can also be extremely creative and can result in new knowledge.

Table 14.2: Summary of RRL spatial analytical research survey

Area of Activity
Fractals
Diffusion modelling
Scale and aggregation
Surface interpolation and generation
Area interpolation
Error modelling
Bayesian methods and poisson regression for rare event modelling
Population estimation
Exploratory geographical analysis
Robust spatial statistics
Frequency domain analysis
Visualization
Integrating statistical packages with GIS packages
Geodemographics and the 1991 Census
Fuzzy spatial data analysis
Supercomputer powered spatial analysis technologies
Artificial intelligence based approaches
Confidentiality issues
Location allocation and network models

Source: E-mail survey of RRLs (April 1989)

Table 14.3: Five key spatial analytical research topics

1	Response modelling for large n data sets with mixed scales and measurement scales or levels
2	Practical methods for cross area estimation
3	Zone design and spatial configuration engineering
4	Exploratory geographical analysis technology
5	Application of Bayesian methods

A further elaboration on the limitations of geographic analysis may be useful. There is also a major difficulty in finding meaningful variables that are free of ecological (and other types of) fallacy. In spatial epidemiology, age/sex standardisation of incidence rates is widespread but neither age nor sex covariates are directly causal variables, they are proxies for other unmeasured and unknown process variables; for instance, age cannot cause cancer but other variables related to age might. So too in a geographical analysis, virtually all the available variables are surrogates and proxies for other variables which are either missing from the database or incapable of measurement or not yet identified. Nevertheless, 'making do' with the available information may identify descriptive relationships that may well be extremely useful as a basis for further research. This suggestive creativity and hypotheses formulating role is nothing to be ashamed about and is a legitimate and extremely worthwhile analysis objective in its own right.

Nevertheless it is important, therefore, to be realistic about what to expect from spatial analysis within a GIS rich environment. Indeed, a number of basic guidelines can be suggested;

(i)   avoid highly formalised scientific designs;
(ii)   adopt an exploratory data analysis mentality;
(iii)   avoid being too statistically blinkered with an over-emphasis on inference;
(iv)   stay within the limitations of geographical analysis; and
(v)   avoid any technique that either implicitly ignores or explicitly removes the effects of space.

GIS is revolutionary technology. It requires flexible fresh thinking, and it needs to be accompanied by a new generation of spatial analytic techniques that can make best use of the new opportunities. Tired, old, wornout, techniques have no place in this brave new world.

The next stage in the argument in favour of EGA is to define a set of basic generic functions. Any such list would have to include:

(i)   pattern spotters and testers;
(ii)   relationship seekers and provers;
(iii)   data simplifiers;
(iv)   edge detectors;
(v)   auto spatial response modellers;
(vi)   fuzzy pattern analysis;
(vii)   visualization enhancers; and
(viii) spatial video analysis.

A little elaboration may be useful. The idea of a pattern spotter is simply an automated means of identifying evidence of geographical pattern in point data sets without any a priori hypothesis to test. This problem may occur in a genuine exploratory situation or as a means of avoiding problems of post hoc model construction when prior knowledge of a specific geographic database renders suspect any hypothesis testing approach. Typically, the objective is that of a spatial search for evidence of pattern; for example, where are the cancer clusters? It is important that the same technology should also be able to test more specific, strictly a priori, pattern hypotheses; for example, is there a cancer cluster near this location? The original concept of a Geographical Analysis Machine (GAM) was an attempt to define a simple special case of a pattern spotter (Openshaw et al. 1987, 1988). Subsequently the concept has been generalised and a family of pattern spotting and testing procedures defined.

A relationship seeker is an attempt to develop a statistical procedure that is a mirror image of the map overlay process. Map overlays can be modelled as a special case of categorical data analysis provided the locational aspects (i.e. the polygon ids) are removed from the cross-tabulation. They could be retained but only at the cost of immense complexity. A simpler more geographically acceptable procedure seems more appropriate. The overlay process can be viewed as a search among $2m-1$ permutations of m map coverages for evidence of spatial pattern being created by the interaction of the overlays with the data of interest. For point data the $2m-1$ overlay permutations can be simulated by the use of virtual maps and a search for the strongest relationships made (Openshaw, Charlton and Cross 1989). The problems of search involving multiple testing is similar to pattern spotting technology and can be solved in a similar manner. A prototype procedure, termed a Geographical Correlates Exploration Machine (GCEM) has been built. An interesting feature is its use of location as an additional level of surrogate variable. It will allow relationships which occur 'here' but 'not there' to be identified. For example, consider a hypothetical disease which is generated by the following process; it is random but with an intensity that depends on people living near an overhead power cable and, either a particular geological structure or near a major road and near a chemical hazard. This combination of 'either/or/and' operations, with surrogate variables would defeat nearly all spatial analysis methods, but GCEM would probably find it. It should not be assumed that spatial patterns are sufficiently simple for eyeballing model residuals to yield good models or that human beings are the best interpreters of complex map overlays. Appropriate automated EGA is needed to seek out the principal spatial relationships contained in complex map patterns.

Data simplifiers are merely GIS relevant versions of existing technology. Regionalisation procedures (i.e. classification with contiguity constraints) provide an obvious means of simplifying very large and complex map databases to identify patterns. Automated zone design procedures provide a neat solution to a whole range of spatial engineering problems; for example, in redistricting and in customised zonal aggregation design. One reaction to modifiable areal unit effects (Openshaw 1984) is to design in an explicit manner spatial data aggregations. GIS is removing all the historic restrictions on the types of zoning systems available for reporting spatial data and it is important that this new found 'freedom' is properly controlled and used.

Edge detection is another area of GIS relevancy that is currently undeveloped. The concept is easily expressed. Zones are usually stored in a vector based GIS as a set of line segments with topological details. Why not develop spatial analytical tools for analysing the data in a GIS at this level instead of at the zonal scale. Patterns may now be seen as a problem in edge detection. An analogous situation in a raster GIS is not difficult to imagine.

Spatial response modelling is another area of considerable importance. Increasingly GIS is creating multiscale databases ranging from the micro level through various stages of macro. There is a need for automated response models to be developed in which the values of a dependent variable can be modelled by reference to whatever spatial predictor information might be available and under circumstances where there is minimal knowledge of the functional forms that might be appropriate. One response is a fully Automated Modelling System (AMS) (Openshaw 1988). Another is to develop variants of the author's Data Base Modelling procedures (DBMs) (Openshaw 1989c). No doubt there are other possibilities, but it is important to remember the design objectives that have been set in this area (i.e. many possible predictors, no prior knowledge of model specification, nonlinearity is to be assumed, data errors would not be unusual, and mixed measurement scales).

Fuzzy analysis procedures are clearly relevant to many areas of GIS. The question is basically how to do it! Currently there are seemingly no operational examples although a fuzzy geodemographic targeting system has been proposed (Openshaw 1989c).

Visualization enhancers represent an attempt to supplement the communication power of the basic static map display. This might be achieved by several routes. A time driven computer movie presentation (i.e. a GIS with video out facilities and basic videotape controller) would enliven an otherwise static display. The quality of the visual images might be further enhanced by performing spatial analysis at each time slice. The effects of space-time analysis might be best seen by displaying a movie of N different, but sequential in time, space-time analyses. For example, the geographic patterning in telephone faults can be displayed at hourly intervals. At each hourly time slice, a simple space-time analysis is performed to determine where patterns occur (as distinct from merely displaying where the faults occur). This type of visualisation enhancement might well provide a simple but extremely effective form of spatial analysis. Quite often it is the simplest tools which turn out to be the most effective. A final area for investigation concerns the possibility of analysing spatial data at the pixel level (Besag 1986). Cellular automata provides one possible tool; another is lattice gauge theory in physics. As processing power becomes less of a constraint, maybe it will become possible to employ image processing technology derivatives on spatial patterns viewed at the pixel scale and by so doing develop a new style of video based spatial analysis technology.

14.6 The future

The future in this area is what spatial analysts want it to be. There are currently few relevant methods and every opportunity to use existing knowledge and skills in statistics, numerical analysis, and computational geography should be used to develop new tools and procedures. The emphasis on new methods reflects the failure of old techniques, the lack of simple spatial analysis problems, poor or inadequate prior knowledge, and a recognition that spatial analysis in a GIS environment is not at all easy or straightforward. Yet it is becoming increasingly urgent and important that the standard range of predominantly map-based data manipulation functions that characterise GIS are complemented by advanced geo-analytic procedures. Many users require now, or soon will, a geographical analysis capability as well as the full range of map manipulative technology. These are exciting times for a spatial analyst interested in GIS applications.

References

Berry, B.J.L., Marble, D.F. (1968) Spatial Analysis, Prentice-Hall, New Jersey

Besag, J.E. (1986) On the statistical analysis of dirty pictures, Journal of the Royal Statistical Society, SocB 48, 192-236

NCGIA (1989) Research Initiatives, NCGIA Update, 1

Openshaw, S., Charlton, M., and Wymer C. (1987) A Mark 1 Geographical Analysis Machine for the automated analysis of point data, International Journal of GIS, 1, 335-343

Openshaw, S., Charlton, M., Craft, A.W., and Birch, J.M. (1988) An investigation of leukaemia clusters by use of a geographical analysis machine, The Lancet, February 6th, 272-273

Openshaw, S., Charlton, M., and Cross, A.E. (1989) A geographical correlates exploration machine', NE.RRL Research Report 3, CURDS, University of Newcastle upon Tyne

Openshaw, S. (1984) The modifiable areal unit problem, CATMOG 38, Geo Abstracts, Norwich

Openshaw, S. (1988) Building an automated modelling system to explore a universe of spatial interaction models, Geographical Analysis, 20, 31-46

Openshaw, S. (1989a) Towards a spatial analysis research strategy for the Regional Research Laboratory initiative, NE.RRL Research Report 1, CURDS, University of Newcastle-upon-Tyne

Openshaw, S. (1989b) Recent developments in Geographical Analysis Machines, NE.RRL Research Report 2, CURDS, University of Newcastle-upon-Tyne

Openshaw, S. (1989c) Making geodemographics more sophisticated, Journal of the Market Research Society, 31, 111-131

Rhind, D., Openshaw, S., and Green, N. (1989) The analysis of geographical data: data rich, technology adequate, Proceedings of the IV International Working Conference on Statistical and Scientific Data Base Management, Rome, Italy, Lecture notes in Computer Science, 339, 425-454, Springer-Verlag

Stan Openshaw
University of Newcastle-upon-Tyne
Department of Geography
Newcastle-upon-Tyne NE1 7RU
United Kingdom

15 GEOGRAPHICAL INFORMATION SYSTEMS AND MODEL BASED ANALYSIS: TOWARDS EFFECTIVE DECISION SUPPORT SYSTEMS

Martin Clarke

15.1 Introduction

Make no mistake, GIS is not only the subject of burgeoning academic interest; it is also big business. It is estimated that in the USA alone, sales of geosystems totalled $282 million dollars in 1988 and it is expected that sales will grow to over $600 million dollars by 1992 (Dataquest Inc. 1989). Government funding of academic research reflects this growth, with the establishment of the National Center for Geographic Information and Analysis (NCGIA) in the States, and the Regional Research Laboratory programme in the UK. GIS has also had the effect of making geography as a discipline 'respectable' - programmes on T.V., articles in newspapers and geographers on government committees have all encouraged a new feeling of self confidence, at least amongst the academic community.

In the face of these developments it would seem churlish to pour cold water on what has been achieved, but it is the central thesis of this argument that, as currently conceived, the progress of the GIS community has stagnated and is failing to achieve a significant fraction of its potential. The reasons that will be offered to support this are centred around the focus that GIS practitioners adopt. I shall contend that the major focus concerns technological issues relating to data storage, retrieval and display. This fetish reflects the type of origins that GIS has - in the defence and remote sensing industries. To be more useful to decision makers, GIS practitioners require a better understanding of the needs of decision-makers in relation to their environments. It will be argued that a crucial requirement for decision makers is the availability of a predictive analytical capability so that managers can begin to explore the possible effects of their decisions. Because the environments and requirements of decision makers vary considerably (in detail if not in substance) the suitability of proprietary GIS software may be called into question viz a viz the merits of customised packages designed with particular clients in mind. The argument therefore centres around the need to move from GIS being an end in itself to becoming an enabling device within a broader decision making environment. This will involve practitioners developing a much broader knowledge of the problems and processes that decision makers are involved in, as well as the incorporation of more powerful, value adding, analytical techniques.

The rest of this chapter is structured as follows. First, we examine some of the main components of conventional GIS packages and relate this to the different system components that GIS packages have been designed to address. We then examine the different functions that GIS packages can or should undertake and the corresponding user-functions in a typical large commercial organisation. This section also describes the importance of value-addedness in GIS and looks at how this is currently being achieved in proprietary GIS software. The conclusion is that, while serving a useful initiating role, the 'analysis' components of most GIS packages are deficient. Section 3 then outlines the types of value-adding methodologies that are available from the geographical modelling community. Examples are given as to how these can be used in practical decision support situations. Section 4 suggests a path towards greater integration between the mainstream GIS community and its modelling counterpart based on an increased sense of awareness of end-user environments and needs.

H. J. Scholten and J. C. H. Stillwell (eds.),
Geographical Information Systems for Urban and Regional Planning, 165–175.
© 1990 *Kluwer Academic Publishers. Printed in the Netherlands.*

15.2 Developments in GIS: Technology led or analysis driven?

Before describing the main features of conventional GIS, it is useful to set the context by defining the main spatial system components that GIS are concerned with. Harris (1989) has usefully distinguished between three main system components:

(i)   The System Container - this is the geographically delimited space that is of interest in a particular application. It may be a country, a region an urban area or a neighbourhood and may itself be subdivided into a number of smaller zones, such as counties, census tracts or postal districts. Computationally, the system container is characterised by a set of spatial relationships (boundary files, locational coordinates, etc.) among a set of defined objects.

(ii) The System Contents - these are the natural and human materials and objects that occupy the system container. Computationally they are characterised by associations with either points (coordinate locations) or areas (such as zones) within the system container. For example, a river is defined by a set of coordinates, a small area population is defined by an association with a zone number.

(iii) System Processes - these are the interactions between the system container and its contents. They may be both physical and social. For example, the journey to work patterns between residential zones and workplaces. Computationally these patterns are more difficult to represent because of the complexity and dimensionality of the processes involved. Often it becomes more efficient to represent them through the use of models which specify the relationships between system contents and their container.

When reviewing the development and applications of GIS we quickly conclude that the main focus has been on (i) and (ii) with little work undertaken on the system process components of spatial systems. However, we can also note that this is precisely the area in which mathematical modelling has been most successful, a point we return to at several stages later in this paper. System container and content analysis as translated in to GIS packages tends to revolve around four main processes - data storage, data retrieval, data query and data display. The reasons for this are quite simple: they reflect the origins of GIS software in the space and defence industries. In these contexts, huge amounts of data, particularly those pertaining to the features of the earth's surface were being collected by satellite. The storage and retrieval of these huge data sets posed severe computational problems and novel solutions to these problems were required. There was then a need to query this database and to extract, say, combinations of features (e.g. altitude and land use) that occurred simultaneously. It was from this that the concept of overlays developed. High definition and quality was also a paramount requirement when these data were displayed. This led to the development of computer mapping technology predominantly on a raster-based approach. Vector and quadtree systems appeared later. Because of the application area there was virtually no concern with system processes such as spatial interaction. The technology developed was then, and remains now, poorly equiped to handle this component. It is this technology that largely underpins modern day GIS packages. The best known, which include ARC/INFO from ESRI, GPPU from Intergraph, SPANS from TYDAC Technologies, TIGER from the US Geological Survey, are all based on a data storage, retrieval, query and display approach. They produce exceptionally pretty maps (usually quite slowly if PC-based) that allow large amounts of data to be interpreted. As such they have important and valid applications in areas where the main concern is with system contents within a container. If one examines these companies' marketing material then most of the applications quoted are in the areas of land use management and in the visual representation of retail market and store centre data. Vendors of GIS systems are quick to deny that they are simply in the business of pretty map production from complex databases. Jack Dangermond, President of ESRI was quoted in the Professional Surveyor recently (1989) as saying "To understand GIS the user has to go through a whole intellectual exercise of developing spatial intellect". But can that intellect be properly developed if it ignores an understanding of the system processes that we described above, facilities for the analysis of which seem lacking in most proprietary GIS packages?

What type of analytical capabilities are represented in proprietary GIS packages? Most GIS systems have the following:

(i)   Arithmetic and geometric analysis: calculate distances between point, areas of defined units, calculate the population within a selected distance from a point, etc.
(ii)   Overlay facilities: a set of Boolean operations that allow points or areas with common set membership to be identified and displayed. For example, identify all areas within 5kms of a motorway, elevation less than 200 metres with existing agricultural land use.
(iii) Statistical analysis: a module that can perform a set of statistical manipulations on system content variables, such as regression analysis.
(iv)   Impact analysis: this is usually restricted to a simple arithmetic or geometric operation, such as calculating and displaying the impact of increased congestion on travel times in the evening peak. Many GIS packages claim much for their impact analysis capabilities and advertise in their marketing literature that it is possible to undertake quite sophisticated forms of analysis. ESRI's claims for ARC/INFO are as follows:

"Another application of GIS is to compare current store locations to competitors' locations using information such as clientele, total annual sales, and product line. This information can then be compared with demographic data to analyze trends, patterns, or anomalies. Once patterns are identified , decision-making becomes easier bacause upper management can quickly understand and react with appropriate responses. ... In this way, researchers can evaluate alternative sites based on the demographics of a community, the proximity of its population to transportation networks and major centres, and the locations of current stores and competitors' stores" (ESRI 1989).

This sounds very impressive, but it is all based on the procedures of data query and overlay. Compared to the type of information a modelling approach would provide it is rather limited. The situation in relation to the current integration of model-based analysis into GIS packages is summarised rather nicely by Figure 15.1 taken from Scholten and Padding (1990).

The fact that the GIS industry has largely failed to take on board the facilities offered by the modelling community reflects a number of trends. First, the main developments in GIS have been technology-driven and based in application areas considerably distanced from the traditional concerns of human systems modelling. The success of GIS thus far is in its technological capabilities rather than its contributions to business planning and analysis. For their part geographic modellers can take little credit. Their failure to draw their potential contributions to the attention of GIS developers reflects an unwillingness to become involved in technology convergence. Indeed, many quantitative geographers have simply switched camps, swapping their failed modeller hats for soon to be failed 'GISer' hats.

In the opinion of this writer there is a need to return to the needs of potential users and forget about the technology for a minute. To recap, for users mainly concerned with system container and system content issues, GIS as presently available provide them with most of what they need. For users who are concerned with system processes (as well as container and contents) then there is almost certainly a need for a 'modelling' capability. It is perhaps useful to draw up a list of activities that managers (at different levels) are concerned with. These (not exhaustively) include:

(i)   routine data query, retrieval and display;
(ii)   intelligence gathering;
(iii) decision support; and
(iv)   strategic planning.

Within an organisation there are what can be described as different user types, which include:

(i)   technology specialists;
(ii)   managers;
(iii) policy makers; and
(iv)   executives.

Figure 15.1: The missing link: the interface of analysis and GIS software

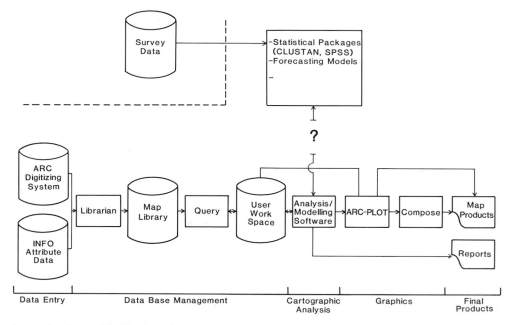

Source: Scholten and Padding (1989)

Figure 15.2 draws up a matrix of the activities against the user groups. The argument is that, currently, GIS use resides in the top left hand corner of this matrix. GIS use typically requires specialists who tend to reside nearer to the base of an organisational pyramid and perform relatively low order functions. What is required is a movement towards the bottom right hand corner where senior staff interface with easy-to-use systems and perform high level tasks.

As we shall continue to argue, these tasks require a significant modelling/predictive ability input.

Figure 15.2: Matrix of typical user groups and activities in organisations

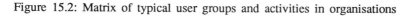

ACTIVITIES

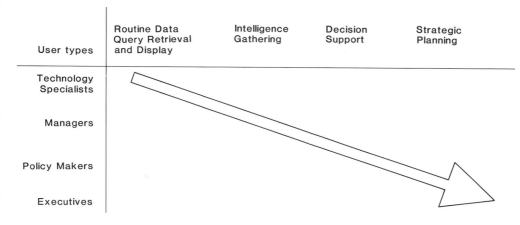

User types	Routine Data Query Retrieval and Display	Intelligence Gathering	Decision Support	Strategic Planning
Technology Specialists				
Managers				
Policy Makers				
Executives				

In many ways there are parallels between the development of geographical information systems and more conventional Management Information Systems (MIS). In many organisations MIS are developed by Data Processing (DP) departments who have as a main interest software and hardware systems. Because they are not directly concerned with the use of the information they produce they tend to see requests for more information as an intrusion on their function and create obstacles to the production of relevant actionable information in a form that is easy for more senior managers to use. In a similar way many of the issues surrounding GIS tend to be of a technical nature and perhaps greater emphasis should be given to how they are to be used in a practical decision support context. In other words there is a need to move GIS technolgy away from technologists and into the hands of decision makers.

15.3 What can models do?

The main use of modelling methods in a practical sense is as tools for adding value to information. This can take a range of different forms which includes:

(i)   providing a framework for data transformation;
(ii)   synthesis and integration of data;
(iii) updating information;
(iv)   forecasting;
(v)   impact analysis; and
(vi)   optimisation.

We describe each of these in turn before returning to the issue of their integration within GIS software.

Providing a framework for data transformation

Models provide a systematic accounting framework that establishes a consistent approach to the analysis of spatial data. This may involve the simple arithmetic manipulation of different items of data. For example market penetration of retail firm x in postal district y can be calculated by summing the expenditure flows from that zone to each store in a region and dividing this by the total expenditure available in the postal district. Plate 15.1 provides an illustration of this in the context of an electrical retailer in West Yorkshire.

In general terms tis type of analysis can add value to data through the generation of performance indicators that can take two fundamentally different types. First, there are those that refer to facilities such as stores, hospitals, schools and so on. In this case we may be interested in the efficiency of a particular specialty within a hospital or the market share of a retailer in a centre. Secondly, there are those that relate to residential areas such as census tracts or postal zones. For example, the variation in hospitalisation rates by postal district may be of concern to a local health authority, as may the levels of market penetration (as discussed above) be of importance to a retailer. The important point to note is that while these may be calculated from existing data, they may also be the product of model calculations as described below.

Synthesis and integration of data

One of the main problems with spatial data is that it generally comes at a wide variety of spatial scales and at different levels of aggregation. For example, the census provides users with a very fine geographical picture of a certain set of variables, but data on income and expenditure is only provided for the standard planning regions of the UK. Models can be used to link and merge different data files and to estimate the values of missing variables. An example of this approach is SYNTHESIS, a package developed at Leeds (Birkin and Clarke 1988, 1989) to generate synthetic micro-data from aggregate probability distributions. For example, using this approach it is possible to generate the spatial distributions of variables, such as income and demographics, at the small area level. Figure 15.3 provides an example of SYNTHESIS output. Here we plot the distribution of dual income no children households ('DINKIES') with joint incomes over 20000 pounds sterling per year in the census wards of Leeds Metropolitan District. The data in this case is derived from two different sources, the 1981 Census of Population and the New Earnings Survey. An alternative approach is to use data compression methods, such as principle components analysis, to condense the wide variety of information available about a spatial unit into a single factor. This approach has been notably successful in the marketing industry with geodemographic products such as ACORN, PIN, MOSAIC, CLARITAS and so on being generated in this way.

Updating information

Information is out of date the minute after it has been collected. Whether this presents a problem to the analyst depends upon the phenomena being studied and the time elapsed since the data were collected. For example, the last population census in Britain was in 1981 and the present day validity of much of the information collected about the characteristics of small areas must be called into question. To provide a solution to the problems posed by such a situation, models can be used to update information. By making best use of a variety of information pertaining to processes such as birth, death and migration, it is possible to generate updated estimates of the characteristics of the population living in small areas, such as census wards or postal sectors. Particularly useful in this context is a method known as microsimulation and examples of its use in the generation of detailed updates of the characteristics of the Leeds population are presented in Table 15.1, taken from a paper describing the 'UPDATE' model (Clarke, Duley and Rees 1989). This tables illustrates how the household composition of one of the city's postal sectors (LS6 1) changes between 1981 and 1986. Note that a considerable amount of disaggregation of household and family type is possible using this particular approach.

Figure 15.3: Distribution of dual income, no children households in Leeds

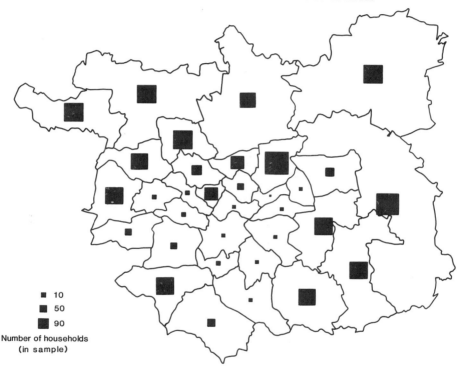

■ 10
■ 50
■ 90

**Number of households
(in sample)**

Forecasting

Whilst updating takes us from the past to the present, forecasting is directed at examining how change may take place in the future. There is a clear requirement for information on future trends in such public services as health, education, social services and so on, as well as in the private sector. It is also important to produce forecasts of population change at a fine geographical scale so that the planning of service distribution meets the specific demands within a region. Models for achieving this type of analysis vary from simple trend extrapolation methods to more sophisticated econometric and mathematical tools. An excellent review of forecasting methods can be found in Wilson and Bennett (1985).

Table 15.1: Household composition tables for a Leeds postal sector, 1981 observed, 1986 estimate

Household Composition	Male Heads					Female Heads				
	16-24	25-34	35-44	45-59	60+	16-24	25-34	35-44	45-59	60+
No Family					1981					
1 person	185	220	75	92	145	133	124	33	82	412
2+ persons	163	159	27	28	40	109	69	17	27	56
					1986					
1 person	22	299	122	90	178	15	169	58	73	381
2+ persons	17	221	56	28	31	13	98	22	25	48
One Family					1981					
MC no children	29	87	27	124	257	6	4	0	3	5
MC dept. only	54	240	101	37	6	11	20	4	2	1
MC non + dept.	0	1	54	77	11	0	0	0	0	0
MC non-dept. only	1	2	4	3	9	7	20	12	13	23
LP dept. only	19	19	13	8	6	50	59	30	13	1
LP non + dept	0	0	3	2	1	0	0	6	13	2
LP non-dept only	0	0	12	73	43	0	0	0	0	0
					1986					
MC no children	0	30	28	74	193	0	5	0	1	3
MC dept. only	0	174	137	40	6	0	17	4	4	2
MC non + dept.	0	18	75	60	15	0	0	3	0	2
MC non-dept only	0	3	15	59	35	0	18	3	15	17
LP dept. only	5	64	13	12	8	9	110	40	14	6
LP non + dept.	1	6	9	7	2	2	15	15	9	6
LP non-dept only	1	3	4	33	64	0	3	19	20	16
Two + Families					1981					
	5	7	5	7	9	3	1	1	2	0
					1986					
	0	9	10	12	18	2	13	4	5	3
All Households					1981					
	456	735	321	451	532	319	297	103	155	506
					1986					
	46	827	469	415	550	41	448	168	166	484

Note: MC = Married Couple;   LP = Lone Parent

Impact analysis

A common issue that models can address is the 'what if..?' question - whether this be the impact of opening a new supermarket, a new hospital or a new car dealership. A range of approaches exist but the key requirement is to be able to estimate the demand for goods or services in small areas, to identify all competing outlets by location and to model the flows of people or expenditure between them. Consider a typical example: a region is divided into 100 demand zones, has 60 shopping centres, and 7 main competitor groups. We are concerned with 5 different product groups and 2 different modes of travel. In all there are 420,000 different possible expenditure flows in this system, and while many of these will be zero, through adopting a model-based approach it is possible to handle this degree of complexity. It also provides the framework described above through which a wide variety of performance indicators can be calculated. For example, in a health care system we could compare performance on a 'before' and 'after' basis. This analysis may show the new plan to be efficient in terms of the facility performance indicators but not effective in terms of residential performance indicators. Figure 15.4 provides an illustration of this type of 'what if..?' analysis. In the first case (a) we illustrate the model estimated revenue of a new supermarket in Leeds on the assumption of all existing supermarkets remaining open. We then assume (b) that a competing supermarket is forced out of business and recalculate the new revenue on this basis. Note that while the pattern of retail expenditure remains similar, the revenue of the new store increases by about 50,000 pounds sterling per week compared with the previous scenario.

Figure 15.4: Impact of new supermarket opening in Leeds

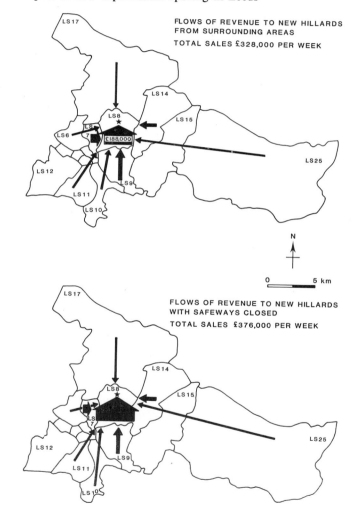

Optimisation

Optimisation methods were first developed under the operations research umbrella, with methods such as linear programming being developed after the Second World War. These methods have been adopted and enhanced for use in spatial analysis, first through approaches such as location-allocation modelling and later through more complex non-linear programming methods (Wilson et al. 1981). In short, these methods are all characterised by the definition of an objective function (e.g. the maximisation of consumer benefits) and a set of constraints, such as minimum levels of supply. The approach demands finding a solution to this problem that maximises the value of the objective function while, at the same time, not infringing any of the constraints. Often such problems are computationally very demanding, for example picking the best 50 locations from, say, 500 possible sites involves examining, in principle 500!/50! alternatives.

These functions that models can perform are almost consistently missing from proprietary GIS software, and where present use very simple and often misleading analysis. In the final section we now address the issue of how the GIS and modelling disciplines can be reconciled.

15.4 A strategy for future development

It may appear from the argument so far that I am painting a picture of the modelling community that is whiter than white, and laying all the blame at the feet of GIS practitioners. This is not what is implied. The modelling community experienced a long period when its tools were largely ignored by commerce, industry and planners. There was a feeling that modelling methods would naturally diffuse into practical application without much assistance from academics - in other words, the demands of users would inevitably lead to model application. This simply did not happen. There were two main reasons for this. The first relates to the ongoing problem - that modellers had little understanding of the problems of 'real world' decision makers and the environments in which they operated. Secondly, the methods, in the form of computer programs, were largley inaccessible, very difficult to use, and very poor in terms of presentation and interpretation of results. The first of these problems can only be overcome through collaborative projects involving researchers working with practitioners and providing tools that better meet their requirements. The second problem began to solve itself through the diffusion of PC systems into both commercial and academic environments. Ironically, it is the presentational facilities that GIS offers that helps make modelling a less bitter pill for the wider community to swallow!

It is therefore argued that the GIS community should learn from this lesson. Understanding the needs of users is paramount to successful GIS use. But what about integrating the types of facilities that model-based analysis offers into GIS software? Well, this will emerge, eventually, through an understanding of user needs, because in several different contexts these can only be met through the application of modelling methods. Whether this happens sooner or later will depend much on the willingness of the two communities to collaborate and to funding agencies to assist this process. There are signs that this is beginning to happen but there is still a long way to go. For this writer, the key factor is recognising the needs of the potential users for information to assist in decision support. If this is achieved then the enormous synergy to be gained from GIS and modelling integration might just be realised.

References

Birkin, M. and Clarke, M. (1988) SYNTHESIS: a synthetic spatial information system. Methods and examples, Environment and Planning, A, 20, 645-71

Birkin, M. and Clarke, M. (1989) The generation of individual and household incomes at the small area level using SYNTHESIS, Regional Studies, forthcoming

Clarke, M. Duley, C. and Rees, P.H. (1989) Micro-simulation models for updating household and individual characteristics in small areas between censuses: demographics and mobility, Paper presented to the International Migration Seminar, Galve, Sweden, (January)

Dataquest Inc. (1989) GIS: the new business opportunity, Business Week, 4/4/89

ESRI (1989) Marketing information for ARC/INFO, ESRI, California

Harris, B. (1989) Integrating a land use modelling capability within a GIS framework, mimeo (copy available from the author)

Scholten, H. and Padding, P. (1990) Working with GIS in a policy environment, Environment and Planning, B

Wilson, A.G, Coelho, J., McGill, S.M. and Williams, H.C.W.L. (1981) Optimisation in Locational and Transport Analysis, John Wiley, Chichester

Martin Clarke
University of Leeds
School of Geography
Leeds LS2 9JT
United Kingdom

16 DECISION SUPPORT AND GEOGRAPHICAL INFORMATION SYSTEMS

Kurt Fedra and Rene F. Reitsma

16.1 Introduction

Geographical Information Systems (GIS) are gaining increasing importance and widespread acceptance as tools for decision support in land, infrastructure, resources, environmental management and spatial analysis, and in urban and regional development planning. GIS assist in the preparation, analysis, display, and management of geographical data. It is in the analysis and display functions that GIS meet Decision Support Systems (DSS). DSS analyse and support decisions through the formal analysis of alternative options, their attributes vis-a-vis evaluation criteria, goals or objectives, and constraints. DSS functions range from information retrieval and display, filtering and pattern recognition, extrapolation, inference and logical comparison, to complex modelling. The use of model-based information and DSS, and in particular of interactive simulation and optimization models that combine traditional modelling approaches with new expert systems techniques of Artificial Intelligence (AI), dynamic computer graphics and geographical information systems, is demonstrated in this chapter with application examples from technological risk assessment, environmental impact analysis, and regional development planning. With the emphasis on an easy-to-understand visual problem representation, using largely symbolic interaction and dynamic images that support understanding and insight, these systems are designed to provide a rich and directly accessible information basis for decision support and planning.

16.2 Decision support

Underlying the concept of DSS in general is the recognition that there is a class of decision problems that is neither well structured nor unambiguous. Such problems cannot be properly solved by a single systems analysis effort or a highly structured computerized decision aid (Fick and Sprague 1980). They are not unique so a one-shot effort would be justified given that the problem is big enough. Neither do they recur frequently enough with sufficient similarity to subject them to rigid mathematical treatment. They are somewhere in between. Due to the mixture of uncertainty in the scientific aspects of the problem, and the subjective and judgmental elements in its socio-political aspects, there is no wholly objective way to find a best solution.

There is no universally accepted definition of DSS. Almost any computer-based system, from database management or information systems via simulation models to mathematical programming or optimization, could conceivably support decisions. The literature on information systems and DSS is overwhelming; approaches range from a rigidly mathematical treatment to applied computer sciences, management sciences, or psychology. Decision support paradigms include predictive models, which give unique answers but with limited accuracy or validity. Scenario analysis relaxes the initial assumptions by making them more conditional, but at the same time more dubious. Normative models prescribe how things should happen, based on some theory, and generally involve optimization or game theory. Alternatively, descriptive or behavioural models supposedly describe things as they are, often with the exploitation of statistical techniques. Most recent assessments of the field, and in particular those concentrating

H. J. Scholten and J. C. H. Stillwell (eds.),
Geographical Information Systems for Urban and Regional Planning, 177–188.
© 1990 *Kluwer Academic Publishers. Printed in the Netherlands.*

on more complex, ill-defined, policy-oriented and strategic problem areas, tend to agree on the importance of interactiveness and the direct involvement of the end user. Direct involvement of the user results in new layers of feedback structures (Figure 16.1). The information system model is based on a sequential structure of analysis and decision support i.e. the relationships shown in the upper part of Figure 16.1. In comparison, the decision support model implies feedbacks from the applications, e.g. communication, negotiation, and bargaining to the information system, scenario generation, and strategic analysis. The realism of formal models is increased, for example, by the introduction of multiattribute utility theory (Keeney and Raiffa 1976, Bell et al. 1977), extensions including uncertainty and stochastic dominance concepts (e.g. Sage and White 1984), by multiobjective, multicriteria optimization methods, and finally by replacing strict optimization, requiring a complete formulation of the problem at the outset, by the concept of satisficing (Wierzbicki 1983).

Figure 16.1: Strategic decision problems: information systems versus DSS approach (partly after Radford 1978)

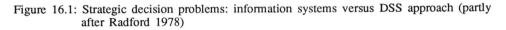

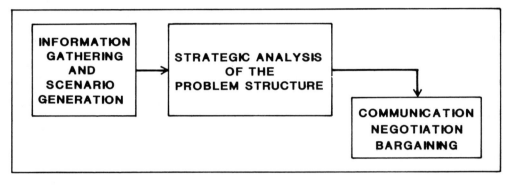

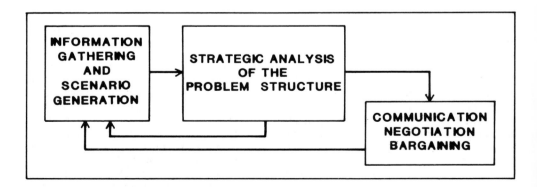

Another basic development is that of getting closer to the users. Interactive models and computer graphics are obvious developments here (e.g. Fedra and Loucks 1985). Decision conferences (Phillips 1984) are another approach, useful mainly in the early stages for the clarification of an issue. While certainly interactive in nature, most methods involve a decision analyst as well as a number of specialists (generally supposed to be the problem holders). Concentrating on the formulation of the decision problem, design and evaluation of alternatives, i.e. the substantive models, are only of marginal importance. Often enough, however, the problem holder (e.g. a regulatory agency) is not specialized in all the component domains of the problem (e.g. regional planning, environmental sciences, etc.). Expertise in the numerous domains touched upon by the problem situation is therefore as much a bottleneck as the structure of the decision problem. Building human expertise and some degree of intelligent judgement into decision supporting software is one of the major objectives of AI. Only recently has the area of expert systems or knowledge engineering emerged as a medium for successful and useful applications of AI techniques (see for example, Pearl et al. 1982, Sage and White 1984, or O'Brian 1985, Zimmermann 1987, Doukidis et al. 1989, on expert systems for decision support). An expert system is a computer program that is supposed to help solve complex real-world problems, in particular, specialized domains (e.g. Barr and Feigenbaum 1982). These systems use large bodies of domain knowledge, i.e. facts, procedures, rules and models, that human experts have collected or developed and found useful to solve problems in their domains.

16.3 Geographical information systems

Geographical information systems are tools to collect, store, retrieve, transform, analyse, and display spatial data. In many applications, it is the automated mapping and cartography, and the basic collection, organization, and management of spatial data that are of primary importance. However, with GIS' capability for analysis and display, and, in a limited sense, modelling, they can be regarded as a special class of decision support systems. Nevertheless, it is primarily in combination with simulation and optimization models, data bases of non-spatial data, AI, and in particular, expert systems technology, and decision support tools proper, that GIS become an extremely powerful and important building block of modern information technology.

16.4 Application examples: integrated systems

As discussed above, it is the integration of GIS and DSS, in combination with simulation or optimization models, related databases, and expert systems tools, that makes attractive and user-friendly decision support tools for a large spectrum of planning and management problems. Several examples of such systems that have been developed at IIASA over the last few years for application areas such as industrial risk assessment, hazardous substances management, environmental impact analysis and regional development planning are introduced below.

Information Retrieval and Display

Simulation and optimization models and model-based decision support tools tend to have considerable data requirements, which, in any particular application, may be a substantial problem for the effective use of these tools. They are therefore coupled, wherever feasible, to a number of data and information bases that provide not only the necessary background information to the user, but allow for the direct and automatic downloading of the model-relevant data. Obviously, for any problem with a spatial dimension, GIS count among the important information sources. For many problems, simple visualization, i.e. the graphical interpretation and display of the problem's structure and components, is an important aid to the decision maker. Translating complex and often abstract concepts such as risk into pictorial representation, and displaying spatial, temporal, and causal relationships in graphical formats often allows a more intuitive understanding of a problem. Pattern, or 'Gestalt', supporting understanding, emerges from graphical interpretation where narrative or numerical formats simply cannot convey the essential features of a problem situation.

A number of databases have been implemented and integrated with interactive browsing and display programs, problem-specific retrieval mechanisms, graphical display options, or linkage to several simulation models in the systems discussed below. While they use and process output, and in particular maps from standard GIS combined with other databases, they offer only a small but very problem-specific subset of GIS functionality in a restrictive but highly efficient menu-driven environment.

As part of a Europe-wide risk management system (Fedra 1985, Fedra and Otway 1986, Fedra et al. 1987) digital maps were used as a background for data display at various levels of aggregation ranging from the European level to very detailed country maps of, for example, the Netherlands. The contents of various databases used in one or several of the simulation models as well as spatial modelling results, can be viewed as interactively constructed map overlays. They include, for example, at the European level, political boundaries, major settlements, highway and national road networks, major industrial plant locations, chemical storage facilities and major water bodies.

In a risk analysis project for the Dutch government (see later section) where a very detailed set of maps with a basic resolution of 1:25,000 was used, features in addition to land use classes and surface waters include the transportation network with rail, road, and canals, city boundaries, or pipelines and power lines, population density, or weather data (windspeed and direction distributions) from individual weather stations. In total, about 50 attribute codes for different area and line features were used. For the raw digital data of approximately 40 MB (vector maps in Digital Line Graph (DLG) format), a number of map manipulation and display routines were developed that allow the efficient generation of various overlay topics as well as arbitrary zooming into the maps (Plate 16.1).

Simulation models can also use digital maps or satellite images as backgrounds for the display of modelling results, or as a guiding framework for spatial problem descriptions. The maps, however, are only pictorial background and are not exploited as data sources. An example is the interactive groundwater modelling system, based on a 2D finite-element simulator (Fedra and Diersch 1989). Combining a finite-element model for flow and transport problems with an AI-based and symbolic, graphics user interface, the system is designed to allow the easy and efficient use of complex groundwater modelling technology in a problem- rather than model-oriented style. Implemented on a colour-graphics engineering workstation, the system provides a problem manager that allows the selection of site-specific, as well as generic, groundwater problems from problem libraries, or the interactive design of a new problem using a map as a background for the definition of problem geometry and hydrogeological parameters (Plate 16.2). With either satellite imagery (such as LANDSAT or SPOT) or DLG standard vector maps as a background, these problems can be edited and modified and then simulated under interactive user control.

Modelling and decision support for spatial problems

A more complex combination of GIS outputs and features is in the functional coupling of spatial data and spatial simulation and optimization models as part of integrated decision support systems. Here data from the GIS are not only displayed, but used in the model-based analysis, with spatial results again displayed and possibly further analysed and interpreted with GIS components.

Regional industrial structure optimization

Industrial structure can be described as an interconnected and possibly spatially distributed network of complementary and alternative production technologies, their technological and economic properties, resource consumption, investment requirements, and the environmental consequences of production, and of course, location. DSS for industrial development planning (e.g. Fedra, Karhu et al. 1987, Fedra, Li et al. 1987, Zebrowski et al. 1988) are based on an optimization model that describes the behaviour of a given group of industries, under certain assumptions about prices for products, raw materials and labour, and upper and lower limits for certain production capacities or waste products. The industry will maximize its net economic results while meeting the external constraints, by adjusting production technologies and production capacities, resulting in a different product mix with different effects on the environment.

This approach can now be extended to include spatial aspects: different locations in a region or country have different sensitivities to environmental pollution, and to risk. As an example, consider the population density around a production plant, its location in relation to important water bodies used for various water supply purposes, etc. In addition, there are of course other important spatial considerations such as transportation costs, risks, and capacity constraints between the individual locations and the sources or markets for raw materials and products. The availability of the necessary technical infrastructure is also a spatial characteristic. Implemented as part of a case study of regional industrial development in China (Fedra, Li et al. 1987), the model considers a set of major production sites and external markets, and more than 140 alternative technologies. The model simultaneously considers criteria such as net and gross production value, export of key commodities (coal and electricity), production cost, domestic and foreign investment, resource consumption, and wastes generated. Local (i.e. spatially distributed) constraints include the availability of certain technologies at a given site, capacity constraints, resources such as coal, water and electricity, and the available labour force. For a given scenario in terms of technologies available, desired product mix and production levels, targets or constraints on the global objectives, and the local site-specific constraints, the model will produce the optimal selection, spatial allocation, and capacity of technologies, if a feasible solution exists at all.

Technological risk analysis

In a study for the Dutch Ministry for Housing, Physical Planning and the Environment (VROM), the Advanced Computer Applications (ACA) project developed an interactive and graphics-oriented framework and post-processor for the risk assessment package SAFETI (Technica 1984) to facilitate the quick generation, display, evaluation and comparison of policy alternatives and individual scenarios. The SAFETI package is a computer-based system for risk analysis of process plants. Based on the detailed geographical background data, the graphical interface to SAFETI's databases and consequence modelling results allows the display of the raw data such as plant locations, weather data, population distribution as thematic overlays over a basic land use map. Most of these data are also used for the computations of the risk estimates for individual plants or transportation problems along any of the transportation networks of the system (rail, road, canals). Once risk analysis has been performed for a specific process plant, the results are available for graphical display and interpretation. Risk contours can be displayed as transparent overlays on a map of the Netherlands. This map allows arbitrary zooming to provide the appropriate level of detail and resolution for a given decision problem.

Air quality modelling

For the regional to local scale, and for continuous emissions, a Gaussian air quality model for multiple point and area sources was adapted. The implementation example described below was designed and implemented for industrial centres in the People's Republic of China. The model was designed as a post-processor for production or energy scenarios based on coal and the regional industrial structure optimization model described above, assessing environmental impacts at a set of locations (areas up to 50 by 50 kilometres), characterized by a number of industrial point and area sources as well as domestic area sources. It translates emission characteristics for these sources into ambient (SO_2) concentrations for a user-defined weather situation. The model provides information on the feasibility and desirability of a given development scenario in terms of selected environmental impacts as one component of the regional development DSS.

From the GIS and related databases, the model constructs a site-specific data file, that is used for display as well as for the calculations, characterizing for one location (industrial installation or zone) the location of the individual sources as well as the default values of emission characteristics. Where available, a background land use map or satellite data are used. On the same grid, elevation data (e.g. from a digital elevation model) used for the dispersion model is stored. The model interface lists the point and area sources and displays a background map of the area studied with the location of the sources indicated. Model results are shown as a colour-coded pollutant concentration overlay on this map, a histogram (using the same colour code) of the frequency distribution of concentration values, or as a 3D concentration field displayed over the tilted and rotated image as illustrated in Plate 16.3.

Transportation risk analysis

In collaboration with INRETS-DERA, France, an information and decision support system for transportation risk assessment of dangerous goods was developed. The system improves availability and fast access to relevant data and information in adequate form (e.g. as topical maps), to support decisions on route selection or modes of transportation. Two kinds of raw data were available for inclusion in the system:

(i)   spatial data: which could be represented as a map, in a vector or raster format, by data on land use, administrative boundaries (communes), road and railway network, and hydrological data; and

(ii) non-spatial data: mostly statistical, which provided the values of attributes of spatial entities (data on the transportation network and its surroundings, commune inventory, and the geographical index for agriculture, as well as data on products and firms concerned with production, usage or shipping of dangerous goods).

The aim was to create a general data structure able to handle these two kinds of data, which could be issued from different sources under different formats, and to connect a geographical object and a statistical object, which was essential for the risk evaluation. The available data was transformed into three different types of data sources accessible from the system:

(i)   spatial information in vector format: data on land use and the road network; the vector format was necessary to allow zooming into the map;
(ii)   spatial information in raster format: data on railways, hydrology and administrative boundaries, stored as rasters; and
(iii) relational database: statistical information was related to the areas, arcs and nodes (e.g. population data to the communes (areas), data on products to the firms dealing with the product, the firms to areas or nodes, and number of lanes, road type (arcs)). Further, the area affected by an accident on a specific arc had to be provided within a given corridor, in terms of types of land use affected.

The transportation risk is described by the probability of an accident and the population in the affected area derived from the statistical and geographical data. The risk factor depends on the properties stored for the arc considered (number of lanes, regular maintenance, etc.). A path generator calculates the minimum risk path which is highlighted on the basic map on the screen. At the same time the user is provided with detailed information from the database on each arc used.

REPLACE: interfacing a knowledge-based model with a GIS

Yet another example of how decision support functions can be linked with GIS is the REPLACE (RElational Plant Location and Acquisition Enquiry) system, a module in the Shanxi Province DSS (Fedra, Li et al. 1987). REPLACE is a knowledge based system based on a matching approach. The idea of approaching site suitability as a matching problem is quite old. Earlier ideas developed in the fifties (e.g. Rawstron 1958), were applied in the feasibility studies of the late sixties and early seventies (e.g. Schilling 1968, Pellenbarg et al. 1974) and in some baseline screening techniques for energy facility planning (Hobbs 1984). 'Relational' matching as applied in REPLACE, however, considers more than a simple matching approach. It is based on the methodological assumption that the definition and measurement of a theoretical concept such as 'site suitability', cannot be based on the empirical characteristics of locations, or the empirical similarities of different locations alone. Instead, site suitability is regarded as the result of a matching of the characteristics and interests of an activity (actor), and the properties of the location (object), as the same empirical properties and combinations of properties of locations, can fulfill very different roles for different activities.

Further, even for the same activities, different combinations of locational properties become important if the objectives and characteristics of that activity change and evolve. What REPLACE has achieved is to develop a model-building technique by which these networks of matching and non-matching actor and object properties and objectives can be represented in such a way that once these representations are available, it is possible to assess the site suitability of various locations for various types of activities (Reitsma 1988). A relational matching model can be regarded as a set-theoretic representation of a concept such as site suitability, i.e. as a set of necessary and sufficient conditions in terms of matching activity and locational properties and characteristics. The matching itself then comes down to a processing of an activity-specific rule base and the data for a set of locations stored in a locational data base, the members of which are tested on the locational requirements inferred from the rule-base. Figure 16.2 shows the conceptual scheme of the core REPLACE system. Running the REPLACE system is simple. After determining the locations which have to be part of the matching, and after determining the exact characteristics of the activity, the dimensionality of the problem, and possible generalizations, the matching program processes the associated rule-base, collects the locational requirements, and checks the locations on these requirements.

Figure 16.2: The core matching model

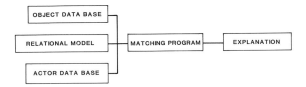

REPLACE and GIS

REPLACE is a typically spatial model. As such, it is not surprising that both from a modelling and a decision support point of view, interaction with a GIS strongly improves the quality of the analysis. GIS increases REPLACE's usefulness in at least three ways. The first and most obvious way concerns the provision of a set of sophisticated mapping functions. In the current implementation of REPLACE, the mapping is done on an ad hoc basis. Several special-purpose maps are stored on disk or generated dynamically. Although this may be sufficient for a first-draft application, it constitutes a rather inflexible way of dealing with the mapping of matching information and matching results. Using the mapping functions of a GIS, on the other hand, could increase this flexibility significantly, and it would relieve the systems programmer from a great deal of unnecessary work which a GIS can do better anyway. A way in which interaction with a GIS strongly increases the quality of the model results is by using GIS spatial analysis functions for the calculation of locational attributes used in the matching. In many instances, locational data for matching is not available on the level of the areas for which it is matched; either because various standard statistics are collected on different spatial-administrative levels, or because the unit of matching does simply not concur with the areas for which information is available. In such instances, information on one level of spatial organization must be converted to another level, something which may require more-or-less complex overlay techniques. GIS are of course particularly good at this type of operation and can therefore be used to conduct them. Note that this does not require direct interaction between REPLACE and a GIS during matching. Proper calculation of locational attributes can be done prior to matching analysis, and the results can be stored in the locational database system.

For the interpretation of results, standard statistical and, spatial information in the form of maps, tables or graphics, is important as a background. For this type of function, a so-called 'dedicated GIS' can prove particularly useful. Such a GIS is labelled 'dedicated' because it contains a set of analysis tools tuned to the particular decision support situation at hand (Reitsma and Makare 1989). In the case of REPLACE, for instance, it was realized that a meaningful interpretation of the matching results could be strongly facilitated by the availability of the means to compare the spatial distributions of the matching results with other locational information. For the GIS component of REPLACE therefore, several tasks were recognized and implemented:

(i)   uni-variate statistical data description;
(ii)   (re)classification of locational attributes;
(iii)   bi-variate statistical analysis;
(iv)   model results must be treated as a variable; and
(v)   mapping on a location-specific basis.

Plate 16.4 shows the screen after the user selects the 'REPLACE data analysis' option from the main REPLACE menu. The screen shows the map of Shanxi Province containing the categorized scores of the attribute displayed on top of the screen, some uni-variate statistics, three frequency histograms, explanation of the colours, and a menu containing further options. Attributes can be selected by choosing the 'Select and display attribute option' from the menu. Three frequency tables are shown: equal interval frequencies, frequencies of the current classification, and cumulative frequencies of the current classification. Attributes are always categorized. Numerical variables are by default categorized in five equal count classes (equal numbers of observations per class). This is reflected in the structure of the two bottom frequency graphs in Plate 16.4. The top frequency graph displays numbers of observations in 50 equal intervals. The bar beneath it corresponds to the current classification. The values associated with the current class boundaries are shown in both the explanation under the map, and the classification frequency graph. In case variables are indeed numerical, uni-variate statistics are displayed. Nominal variables neither need categorization by the system, nor do an equal interval frequency graph and uni-variate statistics apply. These are therefore not displayed. Missing values ('unknown' information) are displayed as white counties on the map.

Re-classification and bi-variate statistics

In order to enable the user to reclassify numerical attributes into a categorization that is more suited to the purposes, a special graphical reclassification routine was developed. Results become available through automatic updating of the map and the frequency graphs, immediately after reclassifying the attribute. Another feature of the GIS component of REPLACE is the opportunity to calculate some bi-variate associations between variables. At present only a simple, two-variable linear regression and a chi-square test for statistical independence have been implemented, but these two do generate some interesting possibilities for the exploration of statistical relations, especially when combined with maps showing the spatial distributions of their variables. Naturally, regression can only be conducted for interval and ratio variables. If regression is requested for a problem containing a nominal or ordinal variable, the system will tell the user that that is not a desirable option. Whereas regression can be used for numeric attributes, a chi-square test for statistical independence can be applied to a combination of categorized variables. And since, for mapping reasons, numeric variables are also categorized, they can be included in a chi-square test as well. Statistically there is of course no reason to conduct a chi-square test on two numeric variables, but combinations of a numeric variable with a nominal one can be interesting. The dedicated GIS in REPLACE also offers the opportunity to treat matching results as locational variables. This provides the user with a possibility to inspect model results, not only in terms of how they were derived in the first place (REPLACE provides the user with an explanation of what happened during the matching process), but also by comparing them with the spatial distributions of other variables, existing patterns of activity location, or other model results.

16.5 Conclusion

As the above examples illustrate, there are numerous ways of interlinking GIS with DSS to various degrees whenever a decision-making problem has some spatial dimension. Visualization of a problem's context and structure and its alternative solutions is one of the most powerful components of decision support. Maps are not only a familiar and easy-to-interpret format, but provide an ideal vehicle for the organization of complex spatial information when combined with other symbolic and graphical forms of representations. These symbolic and graphical representations of alternative options can in turn either be retrieved from data bases with or without statistical processing and interpretation, can be generated by appropriate models, or are interactively defined by the user. GIS and DSS have a lot in common, but they also can complement each other in many applications. Merged with simulation and optimization models, and AI/expert systems technology, they are important building blocks for a new generation of useful and usable 'smart' information technology supporting planners, managers and decision makers.

References

Barr, A. and Feigenbaum, E.A. (1982) The Handbook of Artificial Intelligence, Volume II, Pitman, London

Bell, D.E., Keeney, R.L. and Raiffa, H. (eds) (1977) Conflicting Objectives in Decisions, International Series on Applied Systems Analysis, John Wiley

Doukidis, G.I., Land, F. and Miller, G. (1989) Knowledge Based Management Support Systems, Ellis Horwood Ltd, Chichester

Fedra, K. (1985) Advanced Decision-oriented Software for the Management of Hazardous Substances. Part I: Structure and Design, CP-85-18; Part II: A Prototype Demonstration System, CP-86-10, International Institute for Applied Systems Analysis, Laxenburg, Austria

Fedra, K. and Diersch, H-J. (1989) Interactive groundwater modeling: color graphics, ICAD and AI, in Proceedings of the International Symposium on Groundwater Management: Quantity and Quality, Benidorm, Spain (2-5 October)

Fedra, K. and Loucks, D.P. (1985) Interactive computer technology for planning and policy modeling, Water Resources Research, 21(2), 114-122

Fedra, K. and Otway, H. (1986) Advanced Decision-oriented Software for the Management of Hazardous Substances. Part III: Decision Support and Expert Systems: Uses and Users, CP-86-14, International Institute for Applied Systems Analysis, Laxenburg, Austria

Fedra, K., Karhu, M., Rys, T., Skoc, M., Zebrowski, M. and Ziembla, W. (1987) Model-based Decision Support for Industry--Environment Interactions. A Pesticide Industry Example, WP-87-97, International Institute for Applied Systems Analysis, Laxenburg, Austria

Fedra, K., Li, Z., Wang, Z. and Zhao, C. (1987) Expert Systems for Integrated Development: A Case Study of Shanxi Province, The People's Republic of China, SR-87-1, International Institute for Applied Systems Analysis, Laxenburg, Austria

Fedra, K., Weigkricht, E., Winkelbauer, L. (1987) A Hybrid Approach to Information and Decision Support Systems: Hazardous Substances and Industrial Risk Management, RR-87-12, International Institute for Applied Systems, Laxenburg, Austria (Reprinted from Economics and Artificial Intelligence, Pergamon Books Ltd)

Fick, G. and Sprague, R.H., Jr. (eds) (1980) Decision Support Systems: Issues and Challenges, Proceedings of an International Task Force Meeting (June 23-25), IIASA Proceedings Series, Pergamon Press, Oxford

Hobbs, B.F. (1984) Regional energy facility models for power system planning and policy analysis, in B. Lev et al., Analytic Techniques for Energy Planning, Elsevier Science Publishers, B.V. Amsterdam, pp. 53-66

Keeney, R.L. and Raiffa, H. (1976) Decisions with Multiple Objectives: Preferences and Values Tradeoffs, Wiley, New York

O'Brian, W.R. (1985) Developing "Expert Systems": contributions from decision support systems and judgement analysis techniques, R&D Management, 15(4), 293-303

Pearl, J., Leal, A. and Saleh, J. (1982) GODDESS: A Goal Directed Decision Supporting Structuring System, IEEE Trans. Pattern Analysis and Machine Intelligence, PAMI, 4(3), 250-262

Pellenbarg, P., Schuurmans, F. and Wouters, J. (1974) De Ontwikkelmogelijkheden van Medemblik: Proeve van een Feasibility Study. 3. The development possibilities of Medemblik: a feasibility study, Geografische Instituut, Rijks Universiteit Groningen, The Netherlands

Phillips, L. (1984) Decision support for managers, in H. Otway and M. Peltu (eds) The Managerial Challenge of New Office Technology, Butterworths, London, p246
Radford, K.J. (1978) Information Systems for Strategic Decisions, Reston Publishing Co. Inc., VA

Rawstron, E.M. (1958) Three principles of industrial location, Transactions and Papers of the IBG, 25, 132-142

Reitsma, R. and Makare, B. (1989) Integration of model-based decision support and dedicated Geographical Information Systems, Journal of Geographical Information Systems (in preparation)

Reitsma, R.F. (1988) REPLACE: The Application of Relational Methodology in Site Suitability Analysis in the IIASA-ACA Shanxi Province Decision Support System, Paper presented at the Annual Conference of the Institute of British Geographers, Loughborough, UK, (January 5-8)

Sage, A.P. and White, C.C. (1984) ARIADNE: a knowledge-based interactive system for planning and decision support, IEEE Transactions on Systems, Man, and Cybernetics, 14(1), 35-47

Schilling, H. (1968) Standortfaktoren fur die Industrieansiedlung; ein Katalog fur die Regionale und Kommunale Entwicklungspolitik sowie die Standortwahl von Unternehmungen (Siting factors for industrial locations and locational choice by enterprises: a handbook for regional and communal policy making), Osterreichisches Institut fur Raumplanung, Veroffentlichungen Nr.27, Kohlhammer GmbH, Stuttgart

Technica (1984) The SAFETI Package. Computer-based System for Risk Analysis of Process Plant, Vol.I-IV and Appendices I-IV, Technica Ltd., Tavistock Sq., London

Wierzbicki, A. (1983) A mathematical basis for satisficing decision making, Mathematical Modeling USA, 3, 391-405 (also appeared as RR-83-7, International Institute for Applied Systems Analysis, A-2361 Laxenburg, Austria)

Zebrowski, M., Dobrowolski, G., Rys, T., Skocz, M. and Ziembla, W. (1988) Industrial structure optimization: the PDAS model, in K. Fedra (ed) Expert Systems for Integrated Development: A Case Study of Shanxi Province, The People's Republic of China. Final Report Volume I: General System Documentation

Zimmermann, H-J. (1987) Fuzzy Sets, Decision Making and Expert Systems, Kluwer Academic Publishers

Kurt Fedra
Advanced Computer Applications (ACA)
International Institute for Applied Systems Analysis (IIASA)
A-2361 Laxenburg
Austria

Rene F. Reitsma
Katholieke Universiteit Nijmegen
Geografisch Instituut
Thomas van Aquinostraat 5
6500 KD Nijmegen
The Netherlands

PART VI
EDUCATION AND MANAGEMENT

17   EDUCATION IN GEOGRAPHICAL INFORMATION SYSTEMS

Gerard Linden

17.1 Introduction

Within education, curriculum development is the main instrument for translating educational goals into an action plan for the transfer of knowledge. Consequently, a curriculum will reflect the ideas and perspectives of an educational institute on a subject area in relation to the requirements of a target group and the institute's mission (Linden 1988). Curriculum development is an iterative process in which objectives are defined, appropriate approaches to assessment and teaching planned and adjustments made in the light of formative evaluation. Curriculum development, even in an established subject area, is a demanding and cumbersome task. It becomes more complicated with Geographic Information Systems (GIS) since this subject is still in its infancy. Given the fast changing perspectives on technological possibilities and user requirements, the development of new curricula and the upgrading of existing ones should be subject to permanent critical evaluation and discussion. Part of this ongoing discussion should be aimed at the identification of challenges and ways to respond as insight in the field of spatial data handling grows. For instance, a recent development has been the recognition of the need for management and planning in the field of GIS and related digital geographic information production facilities. This reflects growing recognition of the importance of the user in an area previously largely concerned with the technology. Other issues are the concept of integrated systems and its impact on the organization and products offered and the recognition that geographic information is an economic resource. Although there is a fundamental difference between administrative oriented information systems and geographic information systems as the former lack the spatial dimension, the educational requirements for GIS mirror those for Information Technology (Chorley 1987).

There are three main sections of this chapter. The subject area is introduced first by presenting a bird's eye view on GIS as a tool box. Indications are given for a more comprehensive definition of the subject area by introducing concepts of integrated systems and integrated GIS technology as an emerging field. The fundamental role geographic information plays in modern society is discussed in the next section. Education in GIS is, as a consequence, essential to fully exploit geographic information. In the last section some perspectives on the subject area and adhering educational activities are used as an illustration of the range of educational requirements and different responses possible.

17.2 The subject area

For our purposes and concurring with present practice we define a GIS as a system for capturing, storing, checking, integrating, manipulating, analysing and displaying data which are spatially referenced to the earth. This is normally considered to involve a spatially referenced computer database and appropriate applications software. A GIS contains the following major components: a data input subsystem, a data storage and retrieval subsystem, a data manipulation and analysis subsystem and a data reporting subsystem. GIS is used here as a generic term for the whole field so encompassing Land Information Systems (LIS). Following the above definition of GIS, geographic data processing can be depicted as a number of interconnected

H. J. Scholten and J. C. H. Stillwell (eds.),
Geographical Information Systems for Urban and Regional Planning, 191–201.
© 1990 *Kluwer Academic Publishers. Printed in the Netherlands.*

transactions from input via the database and the adhering analysis facilities to output. This model (see Figure 17.1 (Young 1986) for a first outline) illustrates the existence of a generally valid framework which, depending on the application area intended and/or the product(s) wanted, can be specified into different application-specific geographic information systems. Agreement on the content and level of detail of such a model enables a linkage with other important issues about the processing of geographic data such as facility and information management, quality control and user requirements. It also provides a platform for a concerted interdisciplinary approach.

Figure 17.1: The main components of a GIS

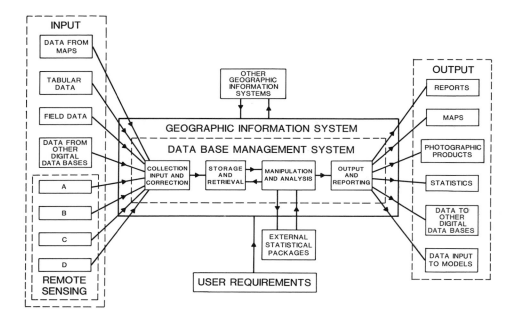

The model depicted in Figure 17.1 shows a quite complete list of input and output facilities with for instance remote sensing as part of the input to a GIS. Remotely sensed data encompass in this model digital data from satellites (A), data from aerial photographs and photographs from space (B), frames from airborne video reconnaissance (C) and data from sensors from other platforms (D). Image processing is understood to be a technique within the field of remote sensing.

A GIS attempts to model the distribution in time and space of natural resources and social and economic indicators. It is in essence a tool box providing the means to access, manipulate and display data to arrive at meaningful information for users such as planners, researchers and administrators. The data may come from many sources such as field surveys, air photography, remote sensing, existing maps and records. A GIS is assessed primarily on its capabilities to analyse geo-referenced data (Burrough 1986). Almost any discipline depending on or making use of information with a spatial component is a potential client for GIS applications. Generally a GIS is capable of performing the following functions: visually and numerically displaying a choice of the data in the database; exploring relationships among spatial data sets; identifying locations which meet specified criteria; supporting decision processes on spatial trade-off problems; estimating impacts through predictive modelling. A GIS as a toolbox offers thus the user flexibility, analysis capabilities by modelling and applying statistics, data integration and presentation options which are far superior to the possibilities of the analogue era.

The above definition of GIS as a toolbox with a rather well defined content, predominantly used for spatial analysis, is no longer covering the subject area adequately. The bulk of GIS technology is not in spatial analysis but in supporting administrative decision making and in the provision of mapping. This introduces also the importance of the organizational context for GIS. Another important development is the emerging concept of an integrated GIS, based on the possibility to receive, store and process data from different sources, for example remote sensing (see also Arnold et al. 1986, and Young 1986). This may take the form of distributed data processing on networked workstations, easy import and export facilities for data from and to other sources into existing packages (compare for instance ARC/INFO) or the design of software with built-in image processing capabilities (ILWIS). The importance of using this concept for the development of the field of geographic data handling can be deduced from a number of recent publications (Rhind 1986, Chorley 1987). For a review of research and development in integrated GIS, see Jackson and Mason (1987). The idea of integrated GIS technology is gaining interest, where, besides the above facilities, additional information such as text, pictures and sound, residing on for instance video based media, may be merged with information residing in a GIS. The subject area of GIS thus contains not only the tool box but also the organizational context in all its aspects and a growing area of related information processing.

17.3 The importance of geo-information

Geographic information encompasses all information which is referenced to the earth and thus has a location. Information can be simply defined in this regard as 'anything that matters' which indicates again the importance of the user who is the one to decide if 'it matters'. Geographic information is not the sole property of one or a couple of disciplines but by character and use, fundamental and instrumental to all disciplines with a spatial dimension. The number of institutes or other interested parties involved in education and training is consequently rather large. Interested parties may be found in the hard and software industry which has to provide training on their equipment, established academic education and research institutes and large organizations with a substantial demand for trained professional staff. Depending on the application, the level and/or the area of competence required will differ.

Traditionally geographic information is displayed in maps, but also in the digital era, (computer) maps still are by far the most important conveyors of geographic information. The main reason is their unrivalled effectiveness in conveying information to a user. Topographical maps on different scales are used for general reference and as the basis for special maps i.e. for urban or rural planning, land suitability analysis or engineering geology. To make these maps geographic data processing is necessary.

This encompasses acquiring, identifying and processing spatially referenced data and then defining, setting up, storing and disseminating the relevant products. Agarwal, Director General of the Survey of India underlines the importance of maps as follows:

"The planners, security forces and law and order enforcement agencies as well as the administrators, economists, agriculturists, geologists and all other working in creative sciences such as engineering etc., find maps to be indispensable aids for any development planning activity, since an overall view of the correct situation existing in respect of resources, the needs of the people, the feasible development alternatives, etc., can be provided by a map. Furthermore, the impact on any development on the human environment etc., can again be monitored and assessed to provide feed-back for corrective measures, only if maps of the areas covered are prepared after implementation of the plan/development schemes. All development activity, whether it be building of a township, road, railway, flood control system or an irrigation system requires a priori detailed information of the topography (physical and cultural) of the area. The ecological complexities of the environment can be assessed only with periodic systematic preparation of maps of the area concerned" (Youngs 1987).

The impacts of the informatics revolution both on developed and developing countries are recognized widely (Cordell 1985, Khadija Haq 1985). Still the importance of geographic information for the management and development of human and natural resources tends to be overlooked and underestimated by its obviousness. This results in a widespread low awareness of the potential contribution of GIS to the management of human and natural resources (Chorley 1987). A large part of a country's development and administrative decision-making has a spatial context. Geographic information is the common denominator of many different activities and thus contributes to the general economy and management of a country. The majority of information processed by, for instance, local and regional authorities, has a spatial dimension. Important application areas of geographic information are amongst others (urban and rural) land use planning, environmental analysis, agriculture and forestry, energy resource development, fisheries and ocean research, market research, real estate cadasters, engineering surveying and design and utility line documentation. (see also Burrough 1986, Arnold et al. 1986, Chorley 1987).

Simple statistics on the availability of up to date maps of different scales in developing countries as compared to the developed world show clearly the enormous backlog in geographic information essential for development (Groot 1987). This observation highlights the need for increased surveying activities on a scale which is only possible with modern technology. Surveying encompasses the inventory of natural and human resources, the assessment of their development potential and the monitoring of change in human activities and environmental processes (ITC 1987a). It is a prerequisite for purposeful action and contributes to decision processes in resource management for conservation and development. The solving of a problem or complex of problems initiates a surveying process to procure relevant geographic information. Depending on the problem(s) to solve and the prevailing constraints, knowledge of the application field dictates the necessary survey design. Surveying is thus not a purpose in itself. The fundamental role of surveying and geographic information for development is important enough to put extra emphasis on education and training as can be illustrated by the following citation:

"The full exploitation of geographic information requires users at all levels to be aware of the relevance and benefits of developments both in general information technology and in geographic information technology. Education is essential to achieving this. Equally important is the existence of trained personnel at all levels, ranging from highly skilled developers of geographic information technology and systems to 'on the ground' operators of systems who need opportunities for training, both 'on the job' and formalized courses" (Chorley 1987).

Information about the land and its characteristics also in relation to the inhabitants - the manner in which they hold land, the settlements in which they reside, and their social and economic characteristics - has become of increasing significance in modern society. The need for better management of (scarce) natural resources be it for development or conservation is especially felt in the rural areas. Change is also obvious in the continued but accelerating pace of urbanization which is expected to show an exponential growth in the remaining years of this century. These observations and the growing acceptance of information as a production factor like capital, machinery and manpower justify the expectation that the demand for spatially referenced information as provided by a GIS will increase (Chorley 1987).

17.4 Perspectives on the subject area: some examples

The emerging awareness of the wider context within which a GIS should be accommodated should be reflected by the GIS education on offer. Still most GIS curricula show a high dependency on training in one or more GIS systems with the emphasis on the various spatial data transactions made possible by the software. Fortunately this rather mechanistic and narrow perspective is being replaced increasingly by a more comprehensive orientation in response to a target group's requirements. This may be illustrated by the following example perspectives of a GIS software vendor, a national academic community interested in proliferating GIS analytical capabilities and an international educational institute involved in more production-oriented GIS applications.

Example of a vendor's perspective: ESRI's PC ARC/INFO University Lab kit

Most vendors of GIS software provide training to their clients in the use of their packages. As the main mission of a vendor is not education and training, introductory courses on specific hard and software tend to be merely 'push button' exercises. This type of training is rightly criticized as it only provides the participants with the 'how' of a specific GIS but does not add to the understanding of general principles and the 'why' of GIS use. An authorative exception is ESRI which offers a PC ARC/INFO University Lab kit especially designed for use in an (academic) educational environment. Apart from the realistic price in view of the eternal university budget problems it also testifies to a well balanced, comprehensive view on the importance of education for the emerging field of GIS and as a consequence ESRI's sincere commitment to the field as such. From a business point of view this long term investment will certainly show to be very profitable. The PC ARC/INFO University Lab kit contains a number of modules, which together with the manuals and a training set provide a complete vehicle for thorough and comprehensive introduction to GIS. For the main modules (Starterkit, Arcplot and Arcedit) a special training module is available encompassing a videotape, a training work book and training data. Although the Lab kit is built around ARC/INFO and its possibilities, the educational platform provided surpasses the product specifications by allowing the teaching of general GIS principles and applications.

Example of an academic perspective: The NCGIA model curriculum

The (US) National Center for Geographic Information and Analysis has developed a model curriculum which should proliferate the analysis capabilities of a GIS (NCGIA 1988). The curriculum is designed as a series of 75 one hour lectures and adhering laboratory exercises divided over three modules. The first module is an introduction to the hardware, software and operations of GIS, providing the essentials required by a beginning GIS technician. The two advanced courses which can be offered after the first module focus respectively on technical aspects and areas related to computer science and computer cartography, and on applied aspects of spatial analysis, spatial decision making and management issues. Course I, the introduction to GIS, covers definitions, examples of applications, map analysis and related technology such as remote sensing. The other subjects are hardware and systems configuration overview, raster-based GIS, data acquisition, the nature of spatial data, spatial objects and relationships, GIS functionality, issues related to raster/vector GIS and trends in GIS. Course II on technical issues, covers coordinate systems and geocoding, data structures and algorithms for respectively vector and raster GIS and for surfaces and volumes, error modelling and data uncertainty, visualization and advanced computational techniques. Course III is on applications of GIS technology. It covers application areas and techniques, decision making in a GIS context, project lifecycle, the multipurpose cadastre, the impact of GIS on management structure, economics and institutions, databases for GIS, data exchange and standards, new directions in GIS and impacts of GIS. The model curriculum is still under discussion and will be tested in a number of educational institutes. The basic driving philosophy in the development of this curriculum is to provide a general education on the basic principles and concepts of GIS, to examine the theory and tools of spatial information analysis and to provide a broad exposure to GIS applications so that objective decisions can be made about system acquisition and implementation.

The model curriculum offers Geography departments and other disciplines concerned with spatial phenomena such as surveying, geology, landscape architecture, forestry and resource management a possibility to integrate GIS concepts in their programs. As may be concluded from the above this perspective and ESRI's perspective are complementary, which renders the PC ARC/INFO Lab kit as natural candidate for hands on training in the laboratory exercises.

The ITC diploma courses in GIS for cadastral, urban and rural applications in developing countries

The majority of subjects included in the NCGIA (university) curriculum may be detected in the curricula of the ITC 11 months full time diploma courses in GIS for cadastral, urban and rural applications. The ITC's mission and the requirements of the target group result, however, in more application-oriented curricula, as will now be elaborated. ITC offers, as other Dutch international educational institutes, courses at a high level. The education system is international in nature, mainly at a post graduate level and intended primarily for participants from developing countries. The emphasis is on mid-career training which shows in the average age of 32 of the participants. Participants for the GIS courses should have a BSc or MSc in a discipline appropriate to their function and adequate work experience on the job. The course should provide them with a thorough understanding of the concepts and techniques of GIS in relation to their respective disciplinary backgrounds.

The GIS cadastral course aims to prepare participants to manage the design, implementation and maintenance of a GIS for cadastral applications. Specific objectives are to enable participants to: establish a cadastral system, national or municipal, for legal, fiscal and other purposes; upgrade or improve cadastral systems; create and develop a municipal management information system; expand an existing cadastral system into a multi-purpose cadaster or large-scale GIS for use with title registration, valuation and assessment, administration and social services, census services, and the development of utilities, services and transportation. A GIS for cadastral applications is a tool for legal, administrative and economic decision-making. It consists, on the one hand, of a database containing spatially referenced land-related data for a defined area and, on the other hand, of procedures and techniques for the systematic collection, updating, processing and distribution of the data. The primary component of a GIS for cadastral applications is a uniform spatial referencing system for the data in the system, which also eases linking data within the system with other land-related data.

The GIS urban course aims at the planning application of urban information systems. Many cities also in the developing world are currently contemplating computerized GIS as a vital element in their strategy to improve quality and control of urban planning and management. The course offers a broad panoramic view to introduce urban technical staff to the variety of possible systems and their potential in relation to eventual applications. Emphasis is on physical urban planning (land use planning, infrastructure planning and housing), but links with other fields of planning (traffic planning, educational planning) land management (cadaster, utility network maintenance) also get ample space in the course. The course aims at active planners in senior technical positions who will be involved in the selection and introduction of computer methods and urban information systems in the planning office and in the urban administration. Upon returning from the course a successful participant can participate in the selection of suitable software and hardware systems and in the definition of database criteria. He (or she) will be able to manage and develop planning and mapping applications of the system(s) that may be eventually installed, in close cooperation with other municipal agencies. He (or she) will understand the potential of data maintained by other agencies and may participate effectively in discussions on organizational and technical aspects of data linkages.

The objective of the GIS rural course is to train (senior) people in the available hard and software for spatial analysis and survey the possibilities of the use of these techniques for resource management (development and conservation). Therefore the course will have elements of computer technical/operational character, elements of resource surveys and elements of management and planning. At the end of the course participants should be able to evaluate which techniques of data collection, processing, presentation, etc are feasible in their own environment. Beside the lectures common for all streams, the following specific rural development aspects will be taught in the common block: economics; agricultural economics; planning theory; land ecology; land evaluation and other subjects depending on the background of the participants.

The ITC MCC course

One example of GIS education is concerned with the requirements of national survey organizations going 'digital'. At the request of the Survey of India (SOI), ITC designed and conducts a three month special course for those officers of the SOI responsible for the implementation of the Modern Cartographic Centre (MCC). This centre is part of the realisation of a more ambitious plan to modernize the SOI. Considering the size of the SOI and the aims of the plan, this undertaking is unprecedented in the world. In line with emerging ITC philosopy the special course for the MCC of the Survey of India is seen as an extension to the regular courses in GIS. The overall framework of the course and its linkages with courses on other topics is illustrated in Figure 17.2. It shows a client-oriented management structure on top of a conglomerate of technology linking diverse disciplines such as photogrammetry and cartography with remote sensing and GIS applications. Within this model A,B,C and D are considered to be principal educational targets (ITC 1987b).

The curriculum design had to take the above into account by giving the participants insight in three major aspects of the management, use and operation of an integrated map and information production facility. The first aspect concerns the linking of data sources to user defined products through the use of integrated digital techniques. The accent is on decision situations facing the person(s) responsible for a specific production line such as map revision and Digital Elevation Models (DEM). The second aspect is about the assessment of the definition of a digital topographical database. The accent is on exposing the participants to the complexity of aspects such as information classification, data structure, qualification of data etc. The third aspect is facility management comprising hard- and software maintenance, documentation and administration, system security, tape management etc.

Figure 17.2: A model of a digital map production unit

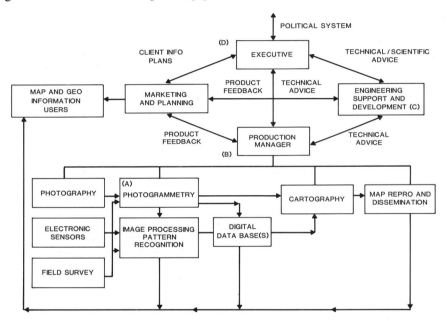

A Unit supervisors day to day production links between units

B Responsible for all aspects of day to day production technology adaption

C Responsible for development of new production capabilities and technical support

D Responsible for inception, justification and execution of programme

Insight into these three aspects should enable the participants to assess the strengths and weaknesses of the system bought. The next step was to divide the available 12 weeks into different subject areas in a logical, effective and practical way. The resulting 5 blocks of different lengths have the following topics: Management Issues on Integrated Digital Map Production (1 week); Photogrammetry, Cartography and Remote Sensing Technology update

(4 weeks); Case Studies of Integrated Production Lines (4 weeks); Data Base Design and Management Issues pertaining to Facility Management (2 weeks) and Testing of Integrated Digital Map Production System (1 week). For each block a detailed programme of lectures and practicals has been worked out on a day to day basis. The emphasis on documentation is primarily necessitated by the demands of evaluation. The top down approach taken to develop the curriculum is evaluated bottom up by weekly feedback sessions with the participants and, following that, intensive discussions between the staff involved. The feedback will be used to improve the curriculum and to design a 12 months MSc level course on the same topic.

17.5 Some concluding remarks

GIS as a subject area encompasses the application of the computer and communication technology to the capture, organization, management and application of spatial information. It deals with all aspects of purposeful spatial database creation, organization, maintenance, access and linkages including the standards and protocols associated with this. The need for such a comprehensive view on the subject area derives from a number of significant characteristics of the information involved, which sets it apart from all other computerized information such as covered by informatics. All data is location-specific. It is very costly to collect and maintain. The amount of data to be stored, transformed and presented is very large and the applications are often unpredictable and extremely varied.

From the above remarks a number of challenges derive. The first challenge for the coming years is to master the concept of an integrated system in all its facets. The introduction of the associated hard and software must serve the need to integrate data inputs from all available sources into digital databases for use in a growing variety of application areas. The second challenge is to include pragmatically in the curricula those elements that bear on socio-economic and social cultural aspects of introducing this new technology in the different production environments. The third challenge is to respond to the recognition that geographic information is an economic resource and that its optimal exploitation will depend on the effectiveness of the infrastructure for access, sharing and value added use of the resource in the development of a country. Awareness of the need for an infrastructure, sensitivity to and realism about user-needs, coupled with the ability to determine for local circumstances the applicability of certain technologies and data sources and the combination of those into practical production lines, are the key assets of personnel who will lead this transition. The fourth challenge is to contribute to the education of this staff and thus strengthen the innovation, production and management capabilities of those institutions. The goal is to increase the contribution to development by including explicitly in curricula and research, issues related to management, quality control, standards and identification of user requirements, as well as the technological subjects per se.

Practicals and training on specialised hard- and software is essential to the curricula. It is almost impossible to teach GIS without hands on experience with relevant exercises on a dedicated GIS package. Present practice shows that the use of simple micro computers to introduce the students to computing and to train them in statistics and text oriented computer applications is successful. The example of the PC ARC/INFO University Lab kit shows that a large part of the training requirements can be met by the use of micro computers and low cost peripherals. For educational, research and consultancy reasons the high end of the GIS hardware and software market should not be neglected. Experience teaches that students of a GIS course should have access to a dedicated GIS for hands on training and general reference during the course.

Theoretically there are many means of assisting the curriculum development process. To be sure that from the start of the curriculum design a comprehensive view of the whole is warranted theoretical models can help to structure the design activities and eventually the evaluation (Knoers and Van der Hoogen 1981). Such a theoretical approach can counter the inclination when evaluating a curriculum to consider and assess the program as a set of separate educational activities. Additional support can be found by using a checklist of curriculum characteristics belonging to such issues as the educational, professional concept, the 'final requirements' of the course, the target group, the structure of the programme and the educational strategy (Chickering et al. 1977, Wardenaar 1984). The process of monitoring and evaluating curricula shows it to be rather cumbersome and time consuming. Improving a curriculum depends largely on a careful design of the evaluation part. As evaluation depends on part of the theoretical basis for curriculum development, an iterative hence time consuming approach is necessary.

A number of other important problems are not discussed explicitly in this chapter. Educating GIS course participants to agents of change has to encompass rising awareness of non-technical factors pertaining to information systems as socio-psychological and organizational issues. In practice the introduction of information systems often fails because these factors are ignored. A recent article in the Dutch professional journal, 'Informatie', revealed this as the real problem of teaching how to design information systems. The authors (Creusen and Mantelaers 1987) stress the importance of knowledge of organisational, sociological and psychological components of the context within which an information system has to be designed. How do organisations function and what role does information play? How do individuals within organisations process information, alternative possibilities for information provision and the interaction between organisation and information?

For the field of geographic information handling, an extra complication is that the context varies with the application, although for some applications, production lines can be detected. The characteristic of GIS, the accent on data manipulation and analysis is fundamental different from the usual business type of information systems. The flexibility created by a GIS is less oriented toward production and is more akin to decision support systems. This demands a type of professional with an equally flexible attitude as opposed to the traditional mechanistic view of information provision within organisations. In this regard it is important to invest in educating persons with different disciplinary backgrounds in the possibilities and constraints of GIS in the various application areas. From the above it is also clear that it will take some time before the full impact of GIS on day to day activities will be felt.

References

Arnold, F., Jackson, M.J. and Ranzinger, M. (eds) (1986) A Study on Integrated Geo-Information Systems, EARSEL Report

Burrough, P.A. (1986) Principles of Geographical Information Systems for Land Resources Assessment, Clarendon Press, Oxford

Chickering, A.W., Halliburton D. et al. (1977) Developing the College Curriculum, Washington

Chorley Committee (1987) Handling Geographic Information, Report to the Secretary of State for the Environment of the Committee of Enquiry into the Handling of Geographic Information chaired by Lord Chorley, Department of the Environment, London

Cordell A. (1985) The Uneasy Eighties; The Transition to an Information Society, Science Council of Canada, Background Study 53, Hull

Creusen, M. and Mantelaers, P. (1987) Teaching the design of information systems: an educational problem, Informatie, 5, 429-438, (original in Dutch)

Groot, R. (1987) Geomatics: a key to country development?, Paper of the Eleventh United Nations Regional Cartography Conference for Asia and the Pacific, Bangkok, (5-16 January)

ITC (1987a) Draft Strategic Plan, ITC, Enschede

ITC (1987b) Provisional Syllabus Special Course for the Modern Cartographic Center of India, ITC, Enschede

Jackson, M. and Mason, D. (1986) The development of integrated Geo-Information Systems, International Journal of Remote Sensing 7 (6), 723-740

Khadija Haq (ed) (1985) The informatics revolution and the developing countries, Paper prepared for the North South Roundtable Consultative Meeting in Scheveningen, The Netherlands, (Sept. 13-15)

Knoers, F. and Van Der Hoogen, J. (1981) How to design a curriculum, University of Nijmegen, Nijmegen

Linden G. (1987) The development of curricula in the field of Geo-Information Systems: the ITC experience, Paper presented at the 2nd International Seminar on Information Systems for Government and Business, UNCRD, Kawasaki

The National Center for Geographic Information and Analysis (NCGIA) (1988) Education Plan, University of California, Santa Barbara

Rhind, D. (1986) Remote sensing, digital mapping and geographical information systems: the creation of national policy in the United Kingdom, Environment and Planning C: Government and Policy, 4, 91-102

Wardenaar, E. (1984) Characteristics of curricula, IOWO, University of Nijmegen, Nijmegen, (original in Dutch)

Young, J.A.T. (1986) A U.K. Geographic Information System for environmental monitoring, resource planning & management capable of integrating & using satellite remotely sensed data, The Remote Sensing Society, Nottingham

Youngs, C.W. (1987) Policies and management of national mapping and charting programmes, Paper presented at the Eleventh United Nations Regional Cartographic Conference for Asia and the Pacific, Bangkok, (5-16 January)

Gerard Linden
International Institute for Aerospace Survey and Earth Sciences (ITC)
550 Boulevard 1945
Postbus 6
7500 AA Enschede
The Netherlands

18 HOW TO COPE WITH GEOGRAPHICAL INFORMATION SYSTEMS IN YOUR ORGANISATION

Jack Dangermond

18.1 Introduction

Hardware and software pose fewer problems today than do data, personnel and institutional arrangements. In this chapter, some of the more important situations in which large, complex GIS must be coped with by an organisation will be indicated and some general suggestions for 'coping' will be offered. These suggestions reflect the experience of the Environmental Systems Research Institute (ESRI), its own work, and the experiences of others with whom ESRI has become familiar over the last twenty years. In this context, coping means solving problems, resolving conflicts, making decisions, taking actions, devising plans, etc. It occurse in response to 'situations'. The word 'situation' is used to avoid the use of a more negative word like 'problem', since in many instances it is hard to say whether a particular situation represents a problem or an opportunity.

This chapter will also describe some processes by which GIS managers can identify their own problems (e.g. the pilot study, the benchmark test, the user needs study, data analysis, and so on) or have their problems identified for them (e.g. using consultants). The chapter generally avoids dealing with those problems of GIS which are independent of the technology itself such as the general problems of management, interpersonal relationships, 'office politics', etc. These are, nevertheless, the most likely causes of ineffective GIS systems.

18.2 Five elements of a GIS

Every GIS is composed of five elements: computer hardware, computer software, data, personnel to run the system, and a set of institutional arrangements to support the other components. Each of these components must be coped with successfully. A GIS is not just hardware and software, or even these two plus data. Hardware and software are probably the least critical elements of the system, in terms of both cost and insuring reliable results. They are closest to being 'off-the-shelf technology': known and reliable quantities. Moreover, most hardware problems will be dealt with by the hardware vendor; software problems by the software vendor.

Data are usually the most costly part of any GIS system (especially if data gathering costs are included), but present mostly technical problems. Data automation problems can sometimes be dealt with by getting an outside service organisation to handle production automation. Then the problem becomes one of quality control and maintaining an assured production schedule. If the decision is made to deal with automation in-house, a major effort will need to be mounted to perform and manage the task; describing that effort is beyond the scope of this chapter. It is the remaining two elements, personnel and institutional arrangements, which are usually the most critical to GIS success.

H. J. Scholten and J. C. H. Stillwell (eds.),
Geographical Information Systems for Urban and Regional Planning, 203–211.
© 1990 *Kluwer Academic Publishers. Printed in the Netherlands.*

18.3 Managing a GIS

To cope successfully, managers need to make use of methods and techniques which tend to flush out problems with the GIS so that they can be identified and deal with. These include:

(i)   user needs studies;
(ii)   decision analysis studies;
(iii)   data analysis;
(iv)   benchmark testing of the GIS;
(v)   pilot studies;
(vi)   implementation planning;
(vii)   project tracking; and
(viii) job costing.

Some of these methods will be discussed below. One way for a manager to cope with many GIS problems is to obtain expert help: from the vendors of the hardware, software, or data; or from expert consultants, either from inside the manager's organisation, or outside it. An increasing number of commercial organisations, professors, and even government employees are working as GIS consultants; some of them may be able to provide helpful advice. A fairly extensive literature exists on how to choose and use consultants effectively.

Successful coping may include some of the following ideas for persons new to GIS technology. First, learn everything you can about how the GIS works. Take training if you can get it. Get briefed, read the documentation and user manuals, observe users, try to get someone to work with you while you are learning the ropes. Try to get to user conferences or meetings of a user group, if one exists. Find someone who is effective and observe her or him; initially, model your approach on this person. Develop a positive attitude; try things; experiment! In order to learn a new technology it is necessary to take risks. Have fun with the technology; the people who use technology most effectively are usually those who enjoy working with it; that enjoyment keeps them going, even when they have problems, and it produces good results.

18.4 Coping with the GIS system cycle

A major factor in determining the amount and character of the coping that an individual or an organisation will have to perform with respect to GIS technology, is where the organisation is with respect to the system cycle of its GIS. For this reason the GIS system cycle will be outlined step-by-step, and the situations which arise and have to coped with, will be descibed. Throughout the system cycle its important to ensure that staff participate in decisions about the GIS, and thus develop a sense of ownership and stake in the success of the system. A major aim should be to gain the cooperation and enthusiastic support of those who will actually use the system.

Preliminary discussions

As soon as a new GIS begins to be discussed, issues of staffing, hiring, budgeting, program responsibility, physical location of the equipment, tasking, reporting, and so on, naturally arise. These may very quickly become matters for dispute. GIS technology can intensify these kinds of disputes because it has a glamour about it, and, when properly used, people recognize that it can have a marked effect on the decision-making process. Hence a GIS can become a centre of power.

GIS technology encourages the sharing of data between diverse agencies and users; thus, a community's planning, engineering, and housing agencies may all want to use the GIS data. This is an obvious virtue of a GIS, but the problem which arises from this is that the agencies which share a GIS may have to agree among themselves and coordinate with one another on details of the system. This need for agreement can be a major stumbling block in getting a GIS into place and may compromise its success. This is why some of the most successful GIS are created by a single agency, using its own funding and under its complete control, with other agencies brought in as users, on some basis, only later. This happens naturally in a fair number of cases, where a single agency has the resources to support such an effort, and goes ahead, independently of other agencies, to develop a GIS. One approach to coping with these situations is to identify the 'champions' of the GIS - those individuals who are prepared to risk themselves and their resources to support the GIS - and support them in their efforts. 'Champions' usually appreciate (even though they may not fully understand) the GIS capabilities, can be persuasive about these capabilities, and have the energy and patience (and sometimes command the resources) to see the whole GIS development process through.

User needs

Assuming that a decision has been made to acquire a GIS, the next step which ought to be taken is an analysis of the needs of users and the decision-making processes which the GIS is intended to influence and be a part of. These surveys are critical because they ask the potential users of the system about what information they need in order to perform their work. Unfortunately, some organisations neglect or avoid this kind of study.

Data analysis

Another element of the careful design of a new GIS is the analysis of existing in-house data, and data otherwise available to the system, in order to determine whether they can support the decision making process which is planned. This analysis leads to decisions about the proposed GIS database, especially the choice of which data will be used. This decision has very large implications for cost, for determining when the GIS will be on-line, for data accuracy and reliability, and so on.

Database design

Database design follows from the results of the user needs study and the data analysis process. This is an area in which professional assistance is particularly appropriate, since so many technical factors are involved, and since the consequences of mistakes are far reaching. Although in the case of some simple database designs, the use of such help might be avoided, having the design reviewed by an outside consultant, or listening to the advice of such a consultant, would probably be wise.

Selecting GIS software

The time has probably passed when an organisation should give any thought at all to writing its own GIS software. The state of the art in GIS technology has advanced so far over the last decade that the cost to duplicate the capabilities of a modern GIS is just too high. While the buyer may feel that all the capabilities of a GIS software system are not needed for a particular application, the reality is that soon after they are acquired, most GIS expand their field of activity very greatly as a result of user requests or demands; so acquiring a system with only limited capabilities is usually a mistake.

Applications software

Buyers need not simply accept the basic features of general purpose GIS software; good GIS software encourages users to customize it, by writing macros for particular user functions, for example.

Drafting the RFP

Given the highly technical nature of the specifications and the system required, consultants may be helpful in drafting the GIS technical specifications, or, if that is done in-house, they might usefully be employed to assist in the process. In specifying system components it is important to recall that the best GIS is not necessarily made up of the best of each kind of component part. Such parts may be incompatible with one another. The system itself must be optimized as a system, not optimized part-by-part.

Specifying hardware

The vast majority of computer hardware used in GIS is general purpose Electronic Data Processing (EDP) hardware, so it is possible to prepare an RFP which will permit many vendors to bid on the hardware for most GIS. Problems can arise in attempting to connect hardware from different vendors. Designers need to know whether the devices are compatible (can 'talk' to one another), or whether various conversion processes or devices will be required. Another problem in specifying the system is how to 'size' the hardware system. Experience here is clear: successful GIS tend to generate greatly increased user demands for capability, so buying extra hardware capacity at the outset is probably a wise investment in most cases. Otherwise, one can expect to have to upgrade a successful GIS within a year or two of its first coming on line.

Specifying software

Software is often 'announced' months or even years before it can be delivered, so that there is a good deal of 'vaporware' on the market: software which is said to exist, and said to possess features, functions, and capabilities which it does not, in fact, yet possess and, indeed, may never possess.

Benchmarks

It is important, in dealing with both 'vaporware' and existing software, that the buyer is able to compare actual performance with promises, by requiring realistic demonstrations of the software's capabilities. The most important mechanism for accomplishing this is the conduct of a benchmark test of the competing software alternatives, preferably on hardware as closely comparable to the buyer's designed system as possible.

Selecting software

Before selecting GIS software, the buyer should probably talk with a number of organisations which use the softwares under consideration, and whose situations are similar to the buyer's own. Buyers should also make every effort to become knowledgeable about GIS software so that they can make the best possible choices among competing products. One problem is that many first-time buyers are unable to distinguish real differences between the capabilities of the various competing softwares.

Visiting user sites

Visits may be useful, but buyers can expect that vendors will tend to direct them to successful GIS sites, and will tend to put the best possible face on things during visits which the vendor supervises. Informal visits and conversations, earlier in the system cycle, may be more useful.

Drafting the RFP

Buyers may be tempted to specify the complete system in minute detail in the RFP, in the hope of obtaining exactly what is wanted. Unfortunately, unless the GIS designer was extremely knowledgeable, the system specified may not function as well as other alternatives which vendors may propose, since vendors usually know their own components better than potential buyers do. Another temptation is to create a checklist of system features and then evaluate vendors chiefly on how many of the features they can supply. Such lists do not, even in sum, create a system design. A GIS is not just a collection of features. Features do not equate to functionality. Some features may be incompatible with one another. Moreover, the features on the checklist will inevitably be of differing importance: if some are missing the effect on the system will be trivial; if others are missing it will be a disaster. A better approach may be to indicate the general capabilities which are needed and then require vendors to describe and justify their best solutions. What is critical is to obtain a GIS which will work, and which will work effectively for the buyer's applications.

Obtaining bids

The next step is to obtain bids from various vendors for either portions of, or for the entire system, and for any needed related services, such as data automation. It is essential to the success of the GIS to reach the relative handful of vendors who can supply the needed system, or the larger group who can supply needed services. There are now so many GIS RFP's out at any one time that no supply organisation is likely to see them all or has time to respond to them all, so some effort by the buyer may be worthwhile to insure that likely successful bidders do respond to the offer to tender. Perhaps some other agencies can provide the names and addresses of suitable firms to which the RFP should be directed. Being aware of what they can provide may be useful in structuring the RFP so as to obtain the needed proposals (ie. not unnecessarily excluding vendors from the competition). The more non-standard a system is, the more difficulty vendors will have in responding to the RFP.

Evaluating proposals

It is important to have enough leeway in evaluating proposals that factors which are most likely to lead to a successful system can be taken into account. This may mean stating them in writing as 'evaluation factors'. A related problem is how each of these should be weighted, and how they can be evaluated in a quantitative way. In evaluating vendors, reputation, past performance, success in the market, availability and quality of training and documentation, and a range of other matters, which are not specifically part of the system design, ought to be considered and appropriately weighted.

Initial cost is obviously an important factor, but the costs of hardware and software will often be the smallest part of the costs of a functioning GIS over its life cycle. Moreover, the costs of a poorly functioning GIS, and of the poor decisions that will flow from it, are likely to be so high (though difficult to evaluate), that obtaining a successful system is probably much more important than low startup costs. Of course, this concept may not be saleable to decision makers or managers on a tight budget. Almost as bad as a complete failure is a situation in which users lack confidence in the system, or find it inconvenient, unresponsive, or inflexible, so that they use it reluctantly. What one is obviously aiming at is a new GIS which will encourage use and provide a significant improvement over the existing system, at an appropriate and reasonable cost.

Benchmark

A means is often needed to determine to what extent the vendor's system actually corresponds to the promises made in the proposal, especially when the capabilities cannot be stated in precise quantitative, measurable terms. Benchmark tests, if they are well designed and executed, are very important, perhaps critical, in making such an evaluation; but conducting such a test is not a trivial task. For small, simple GIS systems the costs involved may not be justifiable. For larger and more expensive systems, or those requiring special features and capabilities, a benchmark is probably one of the best investments that could be made. There are now a few consultants and consulting organisations which can devise and conduct such benchmarks. A list of such vendors can probably be created by contacting recent GIS buyers for whom a benchmark was performed. The fundamental idea of a benchmark is to test the capabilities of rival systems in performing a common set of requirements, thus being able to compare them with one another. This should flush out each system's inabilities, and comparative strengths and weaknesses, in a way that simply reading a proposal cannot usually do. If the buying organisation designs its own benchmark, it may be wise to have vendors comment on it before it is included in the proposal process.

Implementation plan

With all the kinds of information in hand, a buying decision is made and a contract is negotiated. It is important, having selected a GIS, to be realistic and reasonable in devising the implementation schedule for the GIS, given the unknowns which inevitably accompany the acquisition of new technology and its adoption by a user organisation. A portion of the early effort of the selected vendors or consultants should be to actually implement the system, or participate in its implementation along with the buyer's organisation. What is necessary is to plan the implementation process intelligently, to keep the process moving, and then to be flexible. Buyers should be aware that some implementation plans for large or complex systems may require years to complete, especially where large quantities of data have to be automated.

Hardware and software installation

Generally, if the system design was carefully performed, hardware installation should be easy to accomplish and should present few situations in which coping will be required. The same is true for software installation.

User training

User training should have been well thought out and then carefully defined in the contract, with documentation, course schedules, etc., all specified. It should be a standard item from the software seller. Training may be available in several modes, including videotape-based, self-paced instruction. Which of such options is best varies with the buyer's organisation. It is important to be able to train the appropriate number of users, and to insure that managers and decision makers are at least thoroughly briefed on the capabilities and limitations of the new system. Do not economise on this. Ideally, the training should offer some experience with functions very similar to those the organisation will actually use, and should make use of the users' own data and problems.

Pilot study

A pilot study can be extremely useful, perhaps critical, to the success of a new system. It tests and demonstrates a new GIS, helping to flush out those shortcomings, failures, and flaws of system function which could not be foreseen in the design process, and does so before the GIS is fully implemented. The pilot should test all aspects of the system function, on a project for a small pilot study area, which fairly represents the whole of the likely database and the complexity of user functions.

Data automation and updating

Problems with data will not evaporate because one is using a GIS. In some ways they become more acute, because a GIS usually forces more precise data definitions and higher standards of accuracy; errors which go unnoticed on paper maps cannot be ignored in a digital system. Automation may also force an overdue updating of the user's database. The costs involved may be reduced by several strategies. Costs may be shared with other departments, agencies or even outside organisations which would then be allowed to use the same database. These may all be public agencies, or may include public, private and educational institutions. Improved methods of data preparation and standardization may reduce automation costs. Perhaps data which only require conversion can be substituted for data which would have to be captured and automated. Compromising needed data quality in order to speed production is not a good strategy for dealing with this problem. In most cases, high quality data prove to be a very cost effective investment of resources. Initial data automation may be accomplished by a services contract, or a continuing in-house effort. Maintaining the database can be accomplished in the same way or perhaps through transactional updating.

Modelling and analysis

One of the chief reasons for obtaining GIS capabilities is to do modelling. But modelling requires that all the factors which are believed to affect a phenomenon must be represented explicitly, and often quantitatively, in the model of that phenomenon. Modelling thus tends to increase the rationality of any analysis. But modeling also requires that those who make decisions, for example, must be able to explicitly and quantitatively state the reasons for making those decisions; for many kinds of decisions, this is very difficult to do. Even if the criteria can be stated, implementing them in a model may be very difficult, and perhaps misleading. Because the results of the modelling process come from the computer, some may assume that they are more 'objective' or 'quantitative' than other inputs to decision making. But, especially as models become so complex that intuition about them is no longer valid, models can be misleading, or even subject to artful manipulation. So considerable care in modelling and in using modelled results in decision making is required.

New user

It is important that new users share with one another the learning process they are going through, including their successes and their frustrations, and management should make it a point to recognize those methods which are succeeding and use them as models.

Programming

It is possible that users may need programming help from the original software vendor, until the user's staff becomes more proficient with the GIS.

Production work

Initial production may seem, to many novice GIS users and managers, to go too slowly, chiefly because so much care must be taken in the technical processes involved. Dealing with this perception usually requires that it be met head on: yes, the process is slow and meticulous, especially in its early stages, in order to produce a good product. But as new users gain skill and confidence, it will get faster; and once the data are stored in the database, the speed of processing them will increase enormously over manual techniques. It may be as much as a year after a GIS implementation is begun before users really begin to feel comfortable with the new technology and achieve rates of production comparable to those of experienced organisations. If this learning period is unacceptably long, outside help might be gotten to speed the automation process, or do some of the actual production work; or experienced production people might be brought in, for a period of weeks or months, to provide 'on-the-job' training in production; or some staff might spend comparable periods in an existing production shop, learning the most effective production methods.

Coping with a successful GIS

Users, having discovered the benefits of using the new GIS, may begin to overwhelm its capacity to support them. One way to cope is to restrict use of the system, but a better way, often, is to expand or upgrade the system capacity; managers of successful GIS should expect such pressures and have a plan to deal with them. It may be possible to expand on a 'pay as you go' basis as user demand increases. Buying extra hardware capacity to start with is another good way to prepare for this eventuality.

18.5 Continuing the GIS cycle

Software updating

The software system cycle is usually under the control of the GIS software vendor who will periodically provide updates and new releases of software to those who pay the annual maintenance fee. Of course, if one changes vendors or acquires a new software system, this would require a complete analysis of the type already described above.

Hardware updating

Hardware updating is often a more pressing problem than software maintenance. Since the GIS hardware is, for the most part, general purpose EDP equipment, the technology continues to develop and evolve so rapidly that equipment becomes technologically outdated very quickly, as in the shift in the last few years from minicomputers to workstations.

Renewal of the system cycle

Users of GIS technology may be driven to undertake new system cycles at any time. The usual reason is a recognition of additional needs, or new capabilities which need to be taken advantage of, which will justify the trouble and expense of a reiteration of the system cycle. Given today's technology it would be a rare GIS which does not get some kind of overhaul about once every three to five years. In such successive cycles, many of the same problems will arise as in the first cycle; with luck, the problems may be less severe, because some of the people who were around during the first cycle will be available to provide their experience. If skilled people have moved on to other jobs within the organisation, or to other organisations, the lessons learned earlier may not exist in the organisation's memory. This loss of experienced personnel, and experienced users of the technology, is a major problem in many GIS using organisations.

Jack Dangermond
Environmental Systems Research Institute (ESRI)
380 New York Street
Redlands
California 92373
USA

PART VII

DEVELOPMENTS IN HARDWARE AND SOFTWARE

19 GEOPROCESSING AND GEOGRAPHIC INFORMATION SYSTEM HARDWARE AND SOFTWARE: LOOKING TOWARD THE 1990S

Kevin M. Johnston

19.1 Introduction

The past twenty years have seen the development of certain computer-based applications that have changed the way people think about and perform their work. For example, word processing has revolutionised the way text is dealt with. No longer do the majority of people use typewriters; instead they use facilities such as word wrap, letter and word insertion, and automatic centering to process text. Spreadsheets have changed the way in which financial data are stored and manipulated. The manufacture and quality of products have been altered significantly with the advent of Computer-Aided Design and Computer-Aided Manufacturing (CAD/CAM). And GIS are currently modifying our perception of geography. GIS themselves have even begun to experience that ever constant process of change. Indeed, we are now witnessing what might be called the second wave of GIS technology.

This chapter briefly recounts the evolution of computerized spatial analysis, which was first performed with Computer-Aided Design (CAD), then Automated Mapping (AM), Geographic Information Systems (GIS), and most recently, geoprocessing. Because the geographic information system is the cornerstone of many geoprocessing installations, the new data structures of GIS, which feature continuous mapping, single-data storage facilities, and object-based structures, are also looked at and compared with the old systems. In addition, current and future trends in GIS hardware, data acquisition and standards are discussed. Finally, modelling capabilities and legal ramifications are touched upon.

19.2 Historical perspective

Geoprocessing began in the 1960s with the inception of CAD systems. These systems, which perform geometric operations on graphical primitives, proved so effective that cartographers quickly adopted the tools to create maps. The edge of a manufactured item to an engineer and the centre line of a road to a cartographer are represented by the same primitive: a line. Since the intelligence associated with graphical primitives is stored in the mind of the user, professionals from different disciplines can use the same tools to manipulate geometry. But cartographers soon found that the CAD tools could not adequately meet their specialized need for certain spatial representations. That limitation led to the creation of AM, which was continually refined throughout the 1970s. Transformations, map-sheet orientation, and special editing capabilities were some of the essential elements developed in the AM system.

In the late 1970s, the question 'What is the essence of a map?' became a central issue. It was believed that any spatial entity could be represented by three graphical primitives: a point, a line, or a polygon; and by its associated text attributes. Those representations became the basis for the evolution of GIS. The earliest GIS linked a database management system with the graphical tool kits from AM systems. Such linkages allowed users either to query attribute information in a database and see the ensuing graphical representation, or to query the graphical representation and see the attribute information. As users have become more

H. J. Scholten and J. C. H. Stillwell (eds.),
Geographical Information Systems for Urban and Regional Planning, 215–227.
© 1990 *Kluwer Academic Publishers. Printed in the Netherlands.*

sophisticated with old GIS tools, they have found that their representations of the world and its relationships and their modelling techniques are limited. Users have also discovered that additional spatial technologies are needed to assist them in solving problems.

Geoprocessing is the integration of multiple, spatially related technologies: GIS, remote sensing, CAD, surface modelling, civil engineering, facilities design, facilities management, and 3D modelling. With planning problems growing more complex, the integration of multiple technologies will prove to be an invaluable aid to planners in the 1990s.

19.3 Geoprocessing installations and GIS

In many geoprocessing installations (i.e. set-ups involving multiple technologies), GIS are the base technology because they conjoin the database and graphics. Because most of the software technologies, such as surface modelling and remote sensing, which store their data in specialised and restricted formats, use GIS as a common depository, translators have to be created between all the technologies and GIS. The three critical steps involved in establishing a geoprocessing/GIS installation are the acquisition of data, the storage of data, and the utilisation of data, which are the subject of the next three sections of this paper. Geoprocessing is too broad to examine here, so only GIS, the most influential technology in the geoprocessing suite, will be treated. However, the reader should be aware that many of the trends discussed below apply to geoprocessing as a whole.

19.4 Acquisition of GIS data

Experts and users alike have documented that the acquisition and maintenance of a database account for 80 to 90 percent of the cost of establishing a geographic information system. The principal reason for this great expense is that the databases must currently be created by hand digitizing, a process with several flaws. Performed by either a user or a government agency (whose work can be purchased by a user), hand digitizing is so time consuming that the maps it produces are frequently out of date by the time that they have been completed. The process is also neither accurate nor versatile enough to meet all the needs of users working with GIS and other geoprocessing technologies. In the future, data will have to be gathered from more than one source that must be integrated into their respective systems, not gathered and input with a batch translator. In addition, the integrations involved in collecting data will have to occur bidirectionally to maintain the integrity and accuracy of the data.

The two most accurate sources of input for GIS databases are photogrammetry and surveying. The accuracy and value of surveying, which is performed on almost every plot of land, is well known. At present, most paper maps are created by transferring the information gathered by an analytical plotter (the main instrument used in photogrammetry) from its original medium to a digital medium, and then transferring it again to paper or mylar hard copy. Unfortunately, both of the transfers result in the loss of some accuracy. There are also inherent problems in digitizing from hard copy output. The heat from a user's hand, for example, can stretch or shrink a map, confounding the digitizing process. But such difficulties can be minimized by integrating analytical plotters and GIS. Integration allows users to collect their data directly and thus more accurately. With a fully integrated plotter, a user can overlay his existing GIS map on a stereo image (both of which have been corrected for the z-coordinate) and update it accurately.

Raster input too will continue to become important in creating databases. In the 1990s, direct data collectors, such as SPOT, Landsat, and aerial photography, will utilise cameras with exceedingly high resolution. The cameras will need to support stereo imagery because users will no longer be willing to accept the inaccuracies that result from rubber sheeting images and overlaying vector maps (which have been corrected for the z-coordinate). Therefore, raster images will have to be ortho-corrected by the elevation models derived from the stereo imagery.

Even though problems in accuracy will be addressed by improvements in hardware and software, adding intelligence to the entities in an image will remain difficult. Two areas in which techniques for adding intelligence will continue to be developed are image processing and feature recognition.

Raster and vector information will also become truly integrated in the next decade; translators will no longer suffice. Furthermore, raster imagery will be used not only for data collection, but also as backdrops for vector data and for various grid analyses and the scanning of existing maps for data collection and archival purposes. The difficulties in adding intelligence will also exist for the scanned maps, but with slight differences. Feature extractions, for instance, will have to distinguish text from coffee stains and wrinkles, which must be removed. Many users will employ their scanned-in maps as simple raster backdrops for editing or displays, while others will wish only to store their maps for future use.

19.5 Storage of GIS data

Hardware

In the 1990s, GIS installations will comprise multiple computer systems from various hardware and software vendors, who will be forced to work within a heterogeneous environment. The 1990s will also see improvements in networking capabilities, with the workstation continuing to dominate the GIS community. In addition, more intelligence will be developed so that networks can meet the needs of the larger databases and CPU intensive operations that will come into existence. Client-server concepts will mature, managing the large volumes of data and maintaining their integrity and security, as well as distributing processing demands intelligently throughout networks. As in previous decades, the 1990s will see a phenomenal increase in the power of computers and a significant reduction in the cost per MIP.

Standards

The increased demand for integrated geoprocessing technologies, multiple GIS, and heterogeneous hardware platforms will make the 1990s the decade of standards. Users will demand communication standards for the exchange and processing of information on multiple platforms. Protocols such as Ethernet, NFS, and NCS will also be highly requested. And vendors will have to comply with industry standard operating systems; no longer will they be able to dictate proprietary operating systems to users.

The need to pass information from one geoprocessing technology to another so that it is available to the next professional must be expedited in the 1990s. However, since users will want the look and feel of software packages to be similar, they will request windowing systems and standards such as motif and Open Look (based on X-windows). Finally, an increase in the use of multiple technologies will create a demand for common data exchange.

GIS database structure

GIS data structures and storage mechanisms are moving into the second phase of their evolutionary cycle. GIS were initially developed by piecing together and adapting various, and sometimes limited, technologies. Today's systems are being developed from scratch whilst drawing on the experience of the past two decades. Developers are attempting to design systems that will meet work-flow requirements, protect the integrity of databases, and represent the world more effectively, all of which will increase modelling and analysis capabilities. Infant in their functionality but advanced in their data structures, the new GIS will carry users comfortably into the 21st century.

The database, as noted earlier, is the single most important and expensive feature of a geographic information system. In older GIS, data was stored in map sheets, which were pieced together artificially through edge matching (Figure 19.1). That approach has several potential problems. For example, when an entity passes from one sheet to another, it is divided into pieces, one piece for every sheet, which can effect the integrity of the database as well as the analyses performed with the database. Integrity can be jeopardised if a user deletes a piece of an entity that has been broken into multiple entities. For instance, if an entity is a road that passes onto three maps, it will be broken into three separate entities. If the resulting segments are 50, 80, and 35 kilometres in length, and if the user wants to see all the roads that are 120 kilometres or more in length, he will receive 'null' as a response to his query.

Figure 19.1: Early GIS divided the world into maps

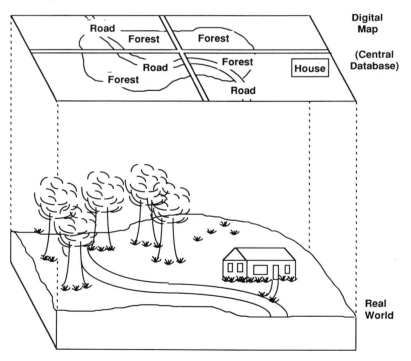

The world does not have artificial boundaries, but is continuous, which is the way new GIS represent it (Figure 19.2). The new systems can support all of the functions of the old systems and more, including allowing users to divide the world into map sheets without compromising the integrity of the database or analyses.

Figure 19.2: New GIS represent the world as being continuous

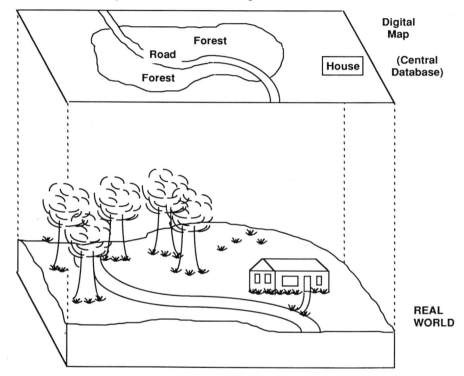

Another area that will witness change in the approaching decade is the extraction of subareas from the main database. Both the old and the new GIS speed up response time by providing users with facilities that enable them to extract subareas for their analyses. Data integrity and analysis are again at risk in the older GIS because the subareas are truncated at the boundary of the area of interest (Figure 19.3). As in the example above, a user can end up deleting the portion of a road that passes through his working subarea without realising that it continues beyond the area he is studying. Or, if a user in search of the minimum environmental criteria for the spotted owl wishes to see all forests that are 50 hectares or greater, he will receive the response of 'null' if he extracts say 35 hectares of a much larger polygon. In contrast, the new GIS allow users access to entire entities that pass through their areas of interest even though they are focused within the map extents of the extracted area (Figure 19.4).

Figure 19.3: Old GIS truncate entities at the boundary of an extraction

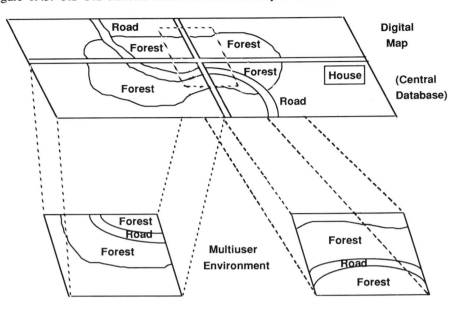

Figure 19.4: New GIS maintain information about the entire entity that passes through an extraction

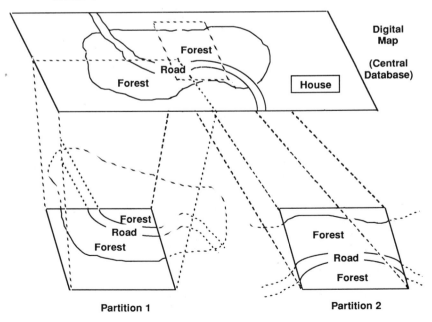

A third area in which old and new GIS differ is the storage of data. The old systems adapt the layer concept of the original CAD, forcing users to divide the world according to homogeneous layers of elements. All forests would be assigned to one layer, for example, and all roads to another (Figure 19.5). One problem with layers is that they create redundancy, which can become a significant problem in large databases. If, for example, the edge of a road also demarcates the edge of a forest, and if forests and roads are stored on different layers, then the edge will be stored twice. Layers can also compromise the integrity of a database. If the centre line of a river is also the boundary of a town district, a voting district, and a county, all of the boundaries may need to move with the river. In old GIS, users have to keep track of which boundary is associated with the centre of the river and then edit each layer accordingly. If they should miss any of the necessary associations, the integrity of the database will be compromised. And since the boundaries of different layers cannot be replicated from one layer to another, overlaying will result in sliver polygons.

Yet another weakness of the old GIS storage techniques is that the graphics and the attributes are kept in different databases (Figure 19.5). Even though the old systems attempt to prevent it, a user can delete an entity from the graphics database without also deleting it from the attributes database and vice versa. Also, the two storage areas do not often allow full functionality to the x, y, and z coordinates, as they do with the attributes.

Figure 19.5: Old GIS divide the world into layers and store attributes and graphics in separate databases

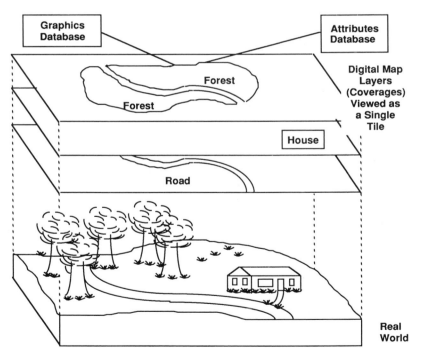

In the new GIS, all entities are stored as a unit (Figure 19.6). Users have the choice of presenting the world either as a whole or in layers. Topology is based on homogeneous characteristics, so nodes do not have to be created at every intersection. But tools do exist that allow a user to share primitives between features or to establish their relationships at any time. If the centre line of the river cited above moves, so would all the other boundaries, if a user arranged for them to. Should the user later decide not to move one of the boundaries, then the previous relationships among the boundaries could be broken. The new systems also store both the graphics and the attributes in a single database, which gives a user full analytical power over the attributes and the coordinates.

Figure 19.6: New GIS represent the world as it is (as a whole) and allow users to define their view of the data

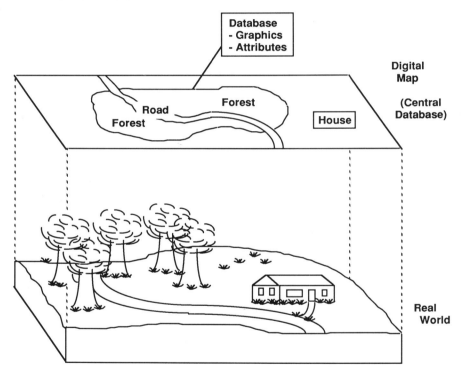

Object- or feature-oriented data structures

Old GIS create entities by first establishing the geometry and building topology with graphical primitives, such as points, lines or polygons. Users can then select a topologic entity and describe it with attributes. The only direct association between entities occurs spatially within a single layer. In the new GIS technology, the creation of entities differs markedly. There, each entity is treated as an autonomous object or feature. Users identify an object first and then describe its geometry, attributes, and relationship to other objects. This object- or feature-oriented approach becomes a powerful tool, not only for producing graphical representations, but also for describing the world and the interaction of its elements.

To illustrate this, let us look at representing a specific entity, in this case a person. Usually a pair of descriptors, such as the name of the person and where he lives, is not enough to provide a user with a complete portrait, especially if the user wants to examine multiple representations of the person, such as his physical appearance, occupation, and marital status. Clearly, a user who knows a great deal about the object he is studying will have a greater understanding and make better decisions concerning the object than a user whose knowledge is limited.

One of the several strengths of an object or feature environment is that it provides an entity with multirepresentations (Figure 19.7), beginning with the raster and the vector representations that exist within the computer. Within the raster and vector divisions, an object can be described in at least nine different ways and used for at least as many different purposes. For example, if the object represented in Figure 19.7 is a parcel of land with a driveway that leads to a building, then vector-text attribute data, such as plot size, ownership, or tax rate can be associated with the parcel and plotted. Scanned raster-text images too, such as the deed to the land, can be associated with the parcel. Spatial correlations do not need to exist between all the representations. A third way of describing the parcel is with a scanned photograph or perspective drawings. If the driveway on the parcel was designed by a civil engineering package or the building by an architectural software package, which can be either raster or vector, then the user can describe the parcel with the information provided by the packages. The user may want to view or do calculations on the elevation of the parcel (see the fifth representation in Figure 19.7), which can be either raster or vector. Yet another way to represent the parcel is through a classical GIS representation (either vector or raster/grid), which shows both the spatial relationships between the entities on the parcel and the parcel's relationship to objects outside itself. The parcel can be also be depicted by direct satellite or aerial photography in which the imagery, once oriented, can be used for display purposes, image processing, or feature recognition. Finally, the parcel can be described by raster images of maps or aerial photographs that can be spatially oriented.

To complete his analysis of the parcel, the user may view one of five distinct scenarios:

(i)   a single raster or a single vector representation of the parcel (e.g. a satellite image or a base vector map;
(ii)   both a raster image and a vector image that are not physically associated (e.g. a vector base map next to a legal description);
(iii)   two physically associated vector images, or a vector-to-vector overlay;
(iv)   a raster-to-raster overlay; or
(v)   a combination of the raster and the vector images, which requires full vector-to-raster integration.

Old GIS attempts multirepresentations through pointers to multiple databases or through multiple technologies in multiple structures. Unfortunately, if a change occurs in one of the representations or if the entity being examined is somehow deleted, the old systems have great difficulty reflecting the change or deletion in the other representations. The object-oriented systems attempt to treat an entity as something in itself with multiple representations. If the entity is changed or deleted, then those alterations will be reflected in all representations. The new systems also store all the representations with the entity itself, which allows the user to make queries about the representations using a single query and a single query language. For example, a user who is attempting to find the best conversion site for a new school, and who wants to see the floor plan of a vacant building that has 10,000 square feet (where square footage is not stored in the GIS data but must be calculated from the floor plan) and stands within three kilometres of major routes, can complete the necessary analysis with a single query in an object-oriented system. In the older systems, many queries in many different query languages (one for each technology) would be required.

Figure 19.7: An object-oriented system allows for multirepresentation of any entity

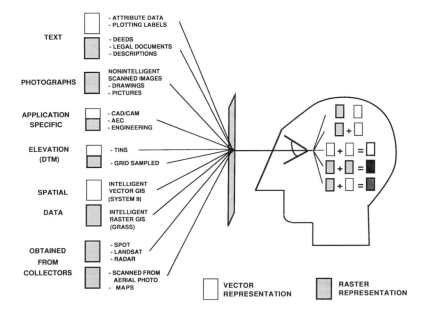

Yet another powerful feature of an object-oriented data structure is its ability to represent the relationships between the entity under observation and other entities in the world. Perhaps the best way to discuss this ability, which can lead to improved analyses and the rapid retrieval of queries, is through an illustration involving a complex entity, in this case, a regional gas network (Figure 19.8). The network consists of a gas-line, clients, and management regions and districts. Its line is composed of main, feeder and customer lines and valves. Each part of the gas-line and the gas-line itself have their own attributes. The customer lines, which directly feed the network's clients, are described by the type of housing that the clients live in: single family dwelling, apartment or condominium. Each type of housing has its own attributes, such as number in household and billing dates. In the new GIS systems, attributes can be assigned to the individual primitives of the network. For instance, cellar depth can be assigned to each primitive that makes up the housing geometry. Customers can be grouped into districts that receive their own attributes, such as individual district managers and building codes. In addition, an association can be made between the gas line and its district that will have its own attributes, including identifying which lines serve which district. And more than one district can be grouped with more than one gas-line to create regions with again, their own attributes, such as individual regional directors, monthly revenues, and number of employees.

Figure 19.8: A complex object model depicting a gas-line network

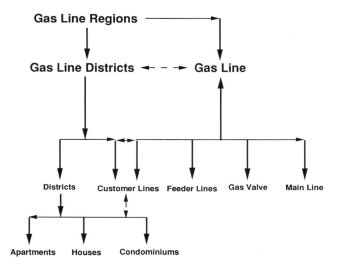

The significance of this array of data, which can be arranged in endless groupings, all with their own attributes, will differ dramatically fron one user to another. A gas maintenance man, for instance, will wish to view the network's stockpile of data differently from a gas delivery man or regional director. The new systems allow for such differing views.

To apply our illustration, let us say that a member of a town maintenance crew accidentally ruptures one of the feeder lines of our network. His crew immediately calls the gas company's home office, where the regional director uses the geometry of his object-oriented GIS to locate the gas leak. The director performs a query to determine in which region and district the leak exists and to learn who the district manager is. He then notifies the manager who then identifies the pipe, its capacity, and the crew responsible for maintaining it, all from a single query. After dispatching the crew, he also simulates the leak on his system to identify which houses are in danger, a step that is facilitated by the fact that his database contains such information as the depths of the cellars in the houses near the ruptured pipe. If necessary, the manager notifies the police so that they can evacuate all residents who are potentially in danger. Through another query, the manager retrieves the capacity of the pipe, its maintenance history, and which valves serve it, information that he then sends to his field crew so that they can fix the pipe.

An involved situation such as a ruptured gas pipe would be handled much less efficiently on an old GIS. First, each component and subcomponent of the network would be placed on separate layers. The only way in which a user could physically associate the layers is through a polygon overlay, a time-consuming process that cannot adequately model the problem. The region involved in our illustration above, for example, would not be composed of a district of houses and the gas-line network; it would consist of only the portions that fall within its geometric extent. Furthermore, the polygon overlay would create inaccuracies in the database. For instance, let us say that a composite overlay has been performed. If the boundary of a district crosses a pipe 50 metres in length, and if that pipe has been divided by the overlay into two pipes of 34 and 16 metres in length, then a user doing a query to identify all pipes longer than 40 metres that need immediate attention will not be aware of the divided pipe. A second problem with forcing users to complete a polygon overlay is that any editing of the original database, which can occur frequently, will necessitate recomputing the overlay.

The new object-oriented GIS allow users to model the world more accurately and to retrieve information more quickly than the old GIS do. The new systems also enable users to create views of the world that will meet their own needs. They provide users with greater database integrity by insuring that the current status of the data will be available at all times; new computations do not have to be performed with each edit. The whole concept of polygon overlay becomes questionable with the new systems, which can reduce the number of times an overlay must be used. Today's users must begin to rethink many of the opinions that the older systems imposed upon them. Modern systems have to be judged on how quickly they can comlete an application, not on how rapidly they perform individual operations.

19.6 Utilisation of GIS data

Modelling

Because GIS technology is still in its infancy, most users have not yet completed building their databases. All of the technology's analytical capabilities have therefore not been exercised rigorously. The 1990s will see increased demands by users for modelling tools. Changes in our world are occurring more rapidly than ever, and each change has a greater influence on the environment than the changes that preceded it. Today's users know that they have to understand problems and their potential consequences and solutions quickly and completely so that they can make appropriate decisions. In the 1990s, users will also request modelling techniques from academics in many disciplines. Statistical analysis will play a prominent role in determining the quality of data and the resultant solutions. And no longer will users be satisfied utilising GIS for only descriptive modelling (i.e. assigning a weighted measure to each area in a study site based on its suitability for supporting the problem at hand). This will also demand that GIS perform prescriptive modelling (i.e. aid them in identifying the best locations to support their problems). During the next decade, GIS will become more than a mapping tool; they will also become an analytical tool.

Legal ramifications

As GIS technology becomes more widespread and databases more useful, the validity and currency of data will be questioned more often. Data analysis too, especially in the area of implementation, will be challenged. And so will the algorithms of the software tools, which are used to perform operations. Certain algorithms actually distort data slightly to perform their functions. Polygon overlay moves points when they fall within certain tolerances of one another, are part of the removal of a silver polgon, or are coincidental. How much movement will be allowed in a law suit involving bodily injury or a fatality? Will rounding errors and inaccuracies of data input hold up in courts of law? The 1990s will undoubtedly see changes in the ways in which GIS are used, especially where the public sector is concerned. And with an increase in liability will come an increase in the cost of databases, for the owner will be forced to pay for liability insurance.

The future of GIS

From the 1960s to the mid-1970s, GIS technology was controlled by the universities. Users, unfamiliar with the technology but aware of its importance, willingly took directions from academics. In the mid-1970s to the present, GIS technology has been taken out of the universities and turned over to the vendors. Again users have followed the prescriptions of a second party. Today's users, however, are better educated than their predecessors were about their tools. They have acquired the experience necessary for understanding what they need and what the current tools are not providing. The 1990s will see geoprocessing technology change from a technology-drive application to a market-driven one. Users now know more than most vendors about how they want to use GIS tools. In the future, end users will tell vendors what they need, and vendors will wisely listen and attempt to satisfy their requests.

19.7 Conclusion

Any technology or field of study abstracts real-world events to make them more comprehensible. Initially, the abstractions are general and the models that represent them vague. But as understanding of the abstractions grows, their models are modified to incorporate the additional insights. The more sophisiticated and flexible the tools for creating models of the real world are, the more readily and comprehensively users will be able to understand the world. And a better understanding of the world can help to ensure that our diminishing resources will be used with greater care.

GIS evolved from the CAD technology of the 1960s. CAD, which was based on fifteen-year-old concepts, could be adapted only so far before its limits were met. The second wave of GIS is now appearing. The newest systems are developed with but one goal in mind; to become GIS. No longer is technology restraining the user from understanding his world. Continuous mapping, extraction of full features, storage of graphics and attributes in a single database, and feature-oriented data structures will lead to improved data integrity and more powerful analytical capabilities. And strengthened data integrity and analyses will mean direct cost savings to end users who invest great sums of money and time in creating their databases and performing analyses. After several rewrites, the word processors of today only slightly resemble the first word processing systems. GIS technology is presently experiencing its first rewrite. Selecting systems that can carry us comfortably into the future has never been of greater importance than now.

Kevin Johnston
Prime/Wild Inc.
Natick
Massachusetts 01760
USA

20 GEOGRAPHICAL INFORMATION SYSTEMS AND VISUALIZATION

Hans E. ten Velden and Max van Lingen

20.1 Introduction

Major changes are likely to take place in the visualization aspects of GIS in the 1990s. Since the time geoscience came into existence, visual output in the form of tables, diagrams, photographs and especially maps has been very important. Over a long period of time the established discipline of cartography has explored the most important means of visualization of geo-referenced and attribute information. The function of visualization however is changing. Over the past decade a variety of computer-based visualization tools have been developed from other disciplines, referred to as 'computer graphics'. Until now the scientific application of those techniques has been limited to the very computation-intensive sciences. But there is now an overall tendency towards integration of different types of software and data, in scientific applications as well as in design, styling and dynamic presentation. There are two main reasons why it is beneficial to look into the world of computer graphics in association with GIS. The first reason concerns the application of the concept of scientific visualization in GIS. The second reason is that in most GIS software packages, graphics are not yet very well developed; there are however possibilities for enhancement of GIS graphics through the use of other software packages.

Since visualization is important in every stage of geo-processing, there is a need for some conceptual framework which captures different forms and functions of visualization related to GIS. The starting-point is to consider visualization as a means of communication at different levels. That results in a proposed multilevel relation between 'data' and 'image', taking both geo-referenced data or the image as a starting-point. It is then possible to functionally link a variety of visualization tools to specific stages of geo-processing.

20.2 Communication through visualization

The visualization of geo-referenced data is in transition. Over a long period of time, maps and charts have been produced by geo-scientists and astronomers. The availability of new computer-based visualization tools makes it both necessary and challenging to reflect on visualisation as an aid in data and information processing in the geo-sciences. Visualization tools such as computer graphics have been developed from different disciplines like applied visual arts, Computer Aided Design (CAD) and modelling. In the past few years the application of computer graphics to fields like computational chemistry, fluid mechanics and geology has revealed the importance of advanced visual aids in scientific research. In the USA, the merging of applied visual arts and science has become a reality. New impulses behind the use of computer graphics in science came from scientists using super-computers. The huge amounts of data produced by complicated computations on super-computers has led to a situation in which, according to the National Science Foundation (NSF), no researcher would be able to interpret thoroughly these piles of data. Using visualization aids, these problems can be overcome. Visual tools can be important in scientific analysis as well as in communication. According to Levine (1988) "it has been estimated that over half the human brain's neurons are devoted to processing and understanding visual input.

H. J. Scholten and J. C. H. Stillwell (eds.),
Geographical Information Systems for Urban and Regional Planning, 229–237.
© 1990 *Kluwer Academic Publishers. Printed in the Netherlands.*

Therefore, to optimize the scientists ability to cope with voluminous sets of numerical data, one must make maximum use of that all-important human visual apparatus. Moreover, one must give the scientist control over those data-driven pictures. Motion and interactivity must be used together. That is the goal of visualization". The main issues in today's debate on scientific visualisation are the technical impediments that prohibit flexible integration of 'V' tools in existing software like GIS. Up to now cost-effectiveness also is a major impediment. Development and integration of tools however, are supported in a recent report of the NSF panel on graphics, image processing and workstations. That report contains a definition of visualization as:

> "a method of computing. In transforms the symbolic into the geometric, enabling researchers to observe their simulations and computations...Visualization embraces both image understanding and image synthesis. That is, visualization is a tool both for interpreting image data fed to the computer, and for generating images from complex multidimensional data sets. It studies those mechanisms in humans and computers which allow them in concert to perceive, use and communicate visual information" (NSF 1987, quoted in the International Journal of Geographic Information Systems 1989).

Visualization can be looked upon as a means of communication at different levels: as man-machine interaction, as a personal analysis tool in a scientific research setting; communication of analysis results between fellow scientists and communication of policy proposals to a general public.

Figure 20.1: Functions of visualization

A. man machine interaction	B. communication to fellow scientists	C. communication to a general public
1. control quality and structure of data	1. producing scientific accountable images	1. explaining results geo-processing activity
2. data analysis	2. producing insight providing images (improving the speed of understanding of compli-cated data-manipulations or computations)	2. promoting policies, ideas, products, etc..
3. (technical) mapping		
4. processing satellite images		
5. real-time simulation		
6. visualize computations		
7. user-interfacing		

Thus the functions of visualization can be divided into three main categories as indicated in Figure 20.1, where each function is illustrated by a series of tasks performed by geo-processing professionals. Note that the difference between the last two categories is mainly the degree to which images are scientifically accountable. For mapping, for example, this means mentioning scale, reliability and so on.

20.3 Current practise

Up to now the possibilities within GIS software to express dimensions like (motion in) time and 3D space in representing data and information are limited because GIS software in general contains only a few graphic modelling features. From the start in the late seventies a typical GIS database structure was two-dimensional, because then the main products of GIS were thematic maps and the 'z' value represented some measured attribute (e.g. number of oak trees per region). Three types of map-producing systems can be distinguished: Computer Aided Drafting and Design (CADD) systems which generate high-quality maps with a data structure consisting of graphic features made of drawing elements or custom symbols organized into layers. AM/FM systems which generate maps often used as data sources for engineering projects involving design, construction and management of facilities like sewer disposal, water or electricity supply; and true GIS which generate maps by analysing and modelling data in a number of ways based on the application (Teichholz 1988). Improvements in GIS map design can be achieved using CADD systems. This combination of software should also facilitate and support more data and analysis-oriented methods of land use design.

In comparison to other disciplines that use visual tools, it may very well be that land use and environmental planners, as an important group of GIS users, have not been able to express their demands very clearly to the software industry. The first applications for example of three-dimensional presentation of geo-referenced data and information in the early eighties are found in surface modelling and technical mapping, not using GIS software (Shoor 1989). Today, tools like real-time 2D and 3D animation, ray-tracing and texture mapping are available and applied in architectural and engineering design and presentation, landscape architecture and image enhancement. For most GIS users, the challenge is to depict the most useful areas for applying these tools. In the RIA project (Scholten and van der Vlugt 1989) of the Dutch National Physical Planning Agency, the concept of 'dynamic cartography' was introduced in a promotional video using 3D animation. Phenomena changing in space and/or in time can be very well presented, while making camera movements around a solid modelled map. This experiment is now taken further by trying to read GIS database files directly into a 3D animation computer, i.e. symbolics. If this method proves successful, there will be more and more presentation of planning proposals and analysis results on video.

The expression and communication of geo-referenced and attribute information in 'still' 2D with cartographic tools like diagram, colour, form, magnitude, structure and scale, widely used in all geo-sciences, will gain extra value from the use of other visual aids. In most cases, means of visualization are used in the last stage of the production of GIS-based maps, as a representation of analysis results. It is possible to use visualization in every stage of GIS data manipulation: in database management (e.g. through visual monitoring of database structures and quality of data), in data retrieval (image driven retrieval), data processing (through image processing or computation) and data representation (dynamic mapping, solid modelling, etc.). In the Orpheus project (Brown 1989), a major technical step was taken of integrating GIS, image processing and CAD software and databases in a hypothetical land use and landscape design project (see Section 20.5). With the introduction in the mid seventies of Computer Aided Mapping (CAM), a facility was created for more rapid and high quality production of maps, using traditional cartographic techniques. But as an analogy with the computational sciences, the visual control over computation and quality of data in terms of accuracy, missing data and reliability may be very useful within GIS database management and various applications. Eventually it may be possible to monitor, in real time, the relation between sudden air pollution and the location and numbers of people affected, for example. In the next section we propose a conceptual framework for interpreting the relation between GIS and visualization, both for scientific and communicative purposes. For illustration of the conceptual framework we draw on the main tasks and products of GIS, a series of articles in Computer Graphics World magazine and a summary of the National Center of Geographic Information and Analysis (NCGIA) research agenda on visualization (International Journal of Geographical Information Systems 1989). In addition, our own experiences in 'dynamic' cartography and animation were used.

20.4 A conceptual framework

What does scientific visualization mean to geo-processing? Taking GIS and visualization as a starting-point we have to consider the main tasks, products, techniques and media for viewing and storage. Besides that we have to look at different types of visualization. The basic elements that are to be considered have been assembled in Figure 20.2. To get a quick impression of relations between these elements one might draw connecting lines between the elements in such a way that it becomes clear that a wide variety of techniques is available for enhancing your GIS product. Present trends towards integration of techniques within spatial analysis as well as visualization will in time facilitate relatively easy-to-use tools. That is not to say that every GIS user will become an expert on visualization. Integration of geo-processing and visualization technology might however stimulate a multidisciplinary approach to planning, design and analysis tasks.

Figure 20.2: Overview of the elements of GIS visualization

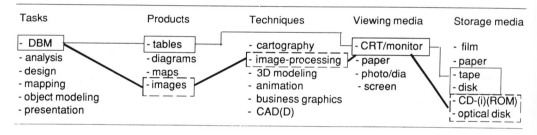

As mentioned earlier, a new approach in visualization can be drawn from those particular techniques that enable us to express motion (in time and space) and facilitate a better man-machine interaction (interactivity). New means of visualization have developed from different disciplines. Animation techniques developed in television research offer dynamic simulation of spatial objects and phenomena. Image-processing facilitates rapid database updating and enhancement of surface models. Interactive media like CD-i and laser vision facilitate the integral storage of large amounts of data, data and object-modelling tools, images and sound. In summary, visual displays can be used to analyse as well as to illustrate information and to generate hypotheses as well as interpret the results of scientific research and communicate them to a general public.

Different types and functions of visualization can be illustrated by an analytic approach to the relation between data and image. The main purpose of these schemes is to stimulate discussion about the role and function of different types of visualization in geo-processing. If we take the data as the starting point the first segment of the relationship between the data and the image is a straight data plot, which may be used as a control on digitizing or as a straightforward output of locational data. No effort is made to enhance or explain the image.

Figure 20.3: Conceptual relations between data and image

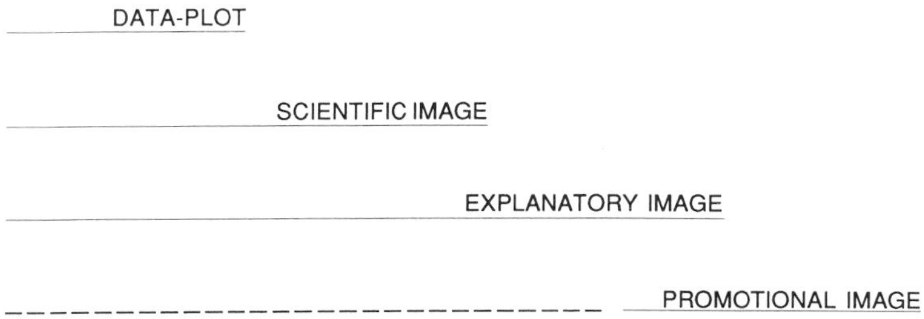

DATA-PLOT

SCIENTIFIC IMAGE

EXPLANATORY IMAGE

PROMOTIONAL IMAGE

In the second segment, images are produced from data and information using visualization techniques that facilitate a better understanding of geo-referenced processes, models, computations, etc. Applied to GIS, it may facilitate a better understanding of the physical condition of the earth or the understanding of socio-spatial processes like migration and spatial phenomena like proximity, contiguity or spatial coincidence. This kind of visualization is referred to as 'scientific visualization'. Data and image are equally important and these visual displays can generate new hypotheses or reveal the quality of the data. Emphasis is on data and/or model representation and interactivity rather then on aesthetic quality. In the third segment, visualization is concerned with communication to research fellows or the general public. Here the long established profession of cartography has a strong tradition e.g. in thematic mapping. Theories of the Frenchman, Bertin, should be mentioned here as Bertin was one of the few people trying to work out systematically which cartographic principle one should use in order to communicate a certain geographic phenomenon or theme to the general public. In the last segment, the direct relationship with the data is very weak. The image is produced through applied visual arts, styling and marketing and is used to communicate results of scientific research, ideas, problems or business proposals so trying to influence perception and/or behaviour.

If we use the image as the starting point (Figure 20.4), the first segment has a weak relation to data. Essentially it is the same as the last stage in Figure 20.3. The second segment represents the position of architectural, industrial or land use design tools, for example. The desired object or land use pattern is designed as an image and from there, a relationship is constructed with attribute data and computational models. Important to note here is the trend in which GIS and CAD are used together as in the Project Orpheus (see section 20.5). The third segment, is user interface design in which the image (on the monitor) serves as an interface between the database and the user. One of the finest examples of user interface design is found in products of the Advanced Computer Applications (ACA) group of the International Institute of Applied Systems Analysis (IIASA) in Laxenburg (Fedra and Reitsma 1989). Using the very powerful multi-tasking options of graphic workstations like SUN, the ACA achieved an very user-friendly interface in which mapping and analysis graphics are combined.

Figure 20.4: Conceptual relations between image and data

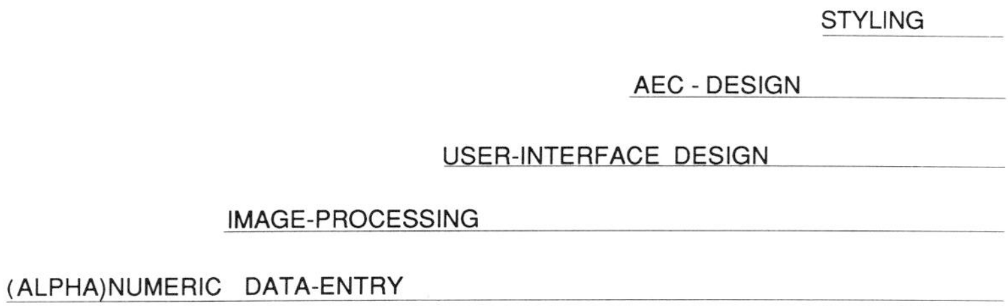

```
                                                    STYLING

                                        AEC - DESIGN

                              USER-INTERFACE  DESIGN

                    IMAGE-PROCESSING

      (ALPHA)NUMERIC  DATA-ENTRY
```

In the fourth segment, images are directly related to database management illustrated by the processing of satellite images used both for scientific interpretation and image enhancement. In the latter case, the image functions as a (pixel) backdrop and can be overlaid with GIS images. Thus the (satellite) image can be transformed to attribute or locational data within the GIS database. Satellite images can also be used to drape over a three dimensional surface model. In the fifth segment the image is no more than a representation of locational data. These conceptual relations serve here as a reference for a global overview of yet available visualization tools that can be used in GIS database management and GIS applications.

20.5 Examples of a new approach to visualization in GIS

Taking the main tasks of geo-processing into consideration and by referring to the proposed conceptual framework in section 20.4, some illustrative applications of visualization tools are briefly mentioned. Some are to be developed and some are available for application in the maintenance and use of GIS.

Database management

For scientific purposes visual monitoring of database structure and data quality (missing data, accuracy, reliability) may well become an important interactive tool. According to the research plan of the NCGIA this is one of the main research areas of scientific visualization in GIS. As yet no formal expressions are available for these applications. When available, it will be possible to produce more scientifically accountable maps for example. Image processing is used for the updating and/or building of geo-databases for example by using a satellite image as a backdrop for an ARC/INFO vector map.

Spatial analysis

The results of statistical analysis of spatial phenomena can be better communicated to a general public using computer graphics. Spatial phenomena both changing in space and time (like migration, transportation or regional economic developement) can be visualized using animation techniques. Dynamic 3D business graphics may illustrate the message more convincingly. Satellite image-processing can be very effectively applied to analysis of land use or environmental pollution.

Spatial design

As mentioned earlier, in land use design a merger is likely to take place between GIS, image processing, CAD and animation software and databases. In a project called Orpheus a hypothetical land use and design problem in the USA was formulated to be entirely processed and supported through a chain of different kinds of software. This experiment was based on the assumption that "land plan problems are far beyond GIS, remote sensing, surface modelling or civil engineering stand-alone solutions" (Shoor 1989). One of the most important implications of the project was "increased data integrity. Instead of dealing with separate collections of possibly conflicting data, the information could be drawn from a single, consistent, uniform database" (Shoor 1989). The project was typically a form of scientific visualization but lacked the integration with animation software which would have allowed for communication of the analysis process to the public. Figure 20.5 shows (future) interconnections as partially used in the Orpheus project.

Figure 20.5: Interrelated technologies and some linkages made using ARC/INFO

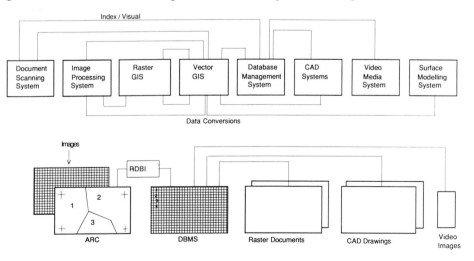

Mapping

New tools applied to map production give the possibility of dynamic mapping. Using 3D modelling and animation software, three-dimensional maps can be produced, showing time-dependent processes. Applications could be found in the modelling and development of groundwater streams and quality, effects of acidification, soil erosion or visualization of socio-economic processes. Most GIS packages produce output that can be digested (through data conversion) by CADD-like or Paint software. For rapid presentations there are several PC-based desktop mapping packages available such as Mapinfo or Atlas*Graphics.

Modelling

Surface modelling can be distinguished from object modelling. Surface modelling is a form of technical mapping and is also referred to as Digital Terrain Modelling (DTM). DTM packages generate three-dimensional representations of the earth's surface. In addition to the impressive graphics, these packages also let users display, analyse and edit ambiguous or statistically problematic data such as faulted surfaces. Techniques used for displaying a third dimesion are triangulation or rectangulation (Brown 1989). Also in this field there is a tendency towards integration. At the moment only a few GIS packages have TIN or RIN modules. ARC/INFO for example, allows surface models to be overlaid with maps of other datasets. Computer aided object modeling is used in architectural, engineering and construction (AEC) software. Together with surface modelling and animation, object modelling is used in advanced flight simulators. These techniques facilitate volume studies in town building for example. Here also a merger with GIS may trigger more indepth study of town and country building.

Presentation

Most techniques mentioned can be used to produce more dynamic and attractive presentations to management or to the general public. Animation and solid modelling as well as texture mapping and image enhancement can be used. Yet most features are rather expensive, but when applied, they will certainly serve the cause.

20.6 Concluding remarks

According to ESRI's Jack Dangermond, "research is sorely lacking in this industry (geo-processing) - we still need to know how users want to use geographic data" (Lang 1988). In our view that statement holds also for visualization and GIS. There are many possibilities to enhance GIS output. Interfaces are being developed between different database formats that facilitate for example the use of Apple computers as a terminal and 'image-processing' tool, combined into a desktop mapping system. In achieving this, database performance is combined with graphic performance. What may be more attractive to GIS researchers is the application of visualization tools within the process of maintenance and manipulation of the GIS database. The proposed conceptual segments of the relationship between the geo-referenced data and the image may serve as a reference to discuss and appoint types of visualization. However, no attempt has been made to construct any formal scientific rule for types of visualization. With new and low-budget visualization software entering the market almost every month, GIS users and application builders have to experiment with different forms of visualization through the integration of geo-processing and computer graphics software and GIS. The software and hardware industry will keep providing their customers with new technology, that means GIS users must express their demands very clearly. Visualization in GIS can be made useful and can be extremely sexy. Of course, it must be seen rather than be written about.

References

Brown, R. (1989) Technical mapping, Computer Graphics World, April, 113-117

Fedra, K. and Reitsma, R. (1989) Decision support and Geographical Information Systems, Paper presented at the GIS Summer Institute, Amsterdam,(14-25 August)

International Journal of Geographic Information Systems (1989) Volume 3, Number 2, 129

Lang, L. (1988) GIS technology puts ESRI on the map, Computer Graphics World, February, 91-94

Levine, R. (1988) Visualization barriers, Computer Graphics World, August, 28-35

Scholten, H. and Meijer, E. (1988) From GIS to RIA, Paper presented at the URSA-Net Conference, Patras, Greece

Shoor, R. (1988) Beyond GIS, Computer Graphics World, February, 87-92

Teichholz, E. (1988) GIS technology matures, Computer Graphics World, December, 56-60

Hans E. ten Velden and Max van Lingen
Rijksplanologische Dienst
Willem Witsenplein 6
2596 BK Den Haag
The Netherlands

PART VIII

INFORMATION BASED SOCIETIES

21 GEOGRAPHICAL INFORMATION SYSTEMS IN PERSPECTIVE

Peter Nijkamp

21.1 Introduction

In the post-war period, many countries have experienced an information explosion. The introduction of computers, micro-electronic equipment and telecommunication services have paved the way for an avalanche of information, not only for scientific research, but also for information transfer to a broader public and for planning or policy purposes (Burch et al. 1979). Several reasons may explain this information explosion in planning and policy-making (Nijkamp and Rietveld 1983, Nijkamp 1988):

(i)   our complex society needs insight into the mechanisms and structures determining intertwined socio-economic, spatial and environmental processes;
(ii)   the high risks and costs of wrong decisions require a careful judgement of all alternative courses of action;
(iii)   the scientific progress in statistical and econometric modelling has led to a clear need for more adequate data and information monitoring;
(iv)   modern computer software and hardware facilities (e.g. decision support systems) have provided the conditions for a quick and flexible treatment of data regarding all aspects of policy analysis; and
(v)   many statistical offices have produced a great deal of data which can be usefully included in appropriate systems.

In recent years, we observe the first signs that micro-electronics, informatics and telematics may dramatically alter western societies. In many countries, prosperity is no longer exclusively created by the production and use of manufactured commodities, but increasingly by the creation and sale of services, notably information-based and knowledge-based services (see also Cordell 1985).

The miniaturization of modern technology due to the development and widespread use of the silicon chip, has already led to drastic changes in transmission patterns of information. Distributed intelligence systems, not only at an intra-firm level but in the future also at a broader scale of European business interactions, are likely to change the face of our societies. CAD/CAM systems focusing on customization and economies of scope are already the predecessors of broader information technologies supporting scientific and economic progress.

Information and communication economics is a recent but rapidly evolving field of scientific research. For instance, several studies show that about half of all economic activity in the USA can be attributed to the processing of knowledge and information rather than of physical goods. This trend has even evoked the question whether conventional sectoral subdivisions (primary, secondary and tertiary) are still very relevant and whether a cross-sectional view on sectors, according to a typology of information intensity, would not be an appropriate complement to our statistical data base, in particular because in almost all countries the information-handling sector of the economy is increasing in importance. Nevertheless, in various countries a widespread use of informatics is still hampered by bottlenecks.

241

H. J. Scholten and J. C. H. Stillwell (eds.),
Geographical Information Systems for Urban and Regional Planning, 241–252.
© 1990 *Kluwer Academic Publishers. Printed in the Netherlands.*

Wilbanks and Lee (1983) mention five bottlenecks precluding a direct and smooth application of information from scientific analysis in policy-making:

(i)   the lack of tailor-made scientific tools for various policy issues, given the time constraints prevailing in policy-making;
(ii)   the discrepancy between basic scientific research and the needs of planners and politicians;
(iii)   the existence of gaps in our knowledge (for instance, interaction effects across disciplinary boundaries, institutional uncertainties, unforeseeable events, etc.);
(iv)   the lack of integration in scientific research, leading to a production of piece-wise kinds of information; and
(v)   the lack of learning from experiences (especially failures) from the past.

It is evident that a user-surveyor communication is necessary for removing the above mentioned bottlenecks. It is of course important that the user or client is not disconnected from an information system, but it is equally important that an analyst is informed about the way a certain policy issue or problem is structured. The modern communication technologies provide no doubt an enormous potential, although these cannot replace the contacts between users and analysts. In several choice situations, however, interactive simulation experiments and computer graphics, designed by experts, can nowadays already directly be used by decision-makers and planners, so that policy and analysis may be brought closer together in the future.

User-oriented information systems are thus necessary in order to improve communications between analysts and policy makers and to avoid a 'black box' view of policy analysis and modelling. The modern communication and information technology offers many perspectives in this regard, as analytical tools can be made more accessible to policy-makers through desktop computer terminals and user-friendly software (interactive computer graphics, for example). Recent developments in adaptive information systems have to be applied in regional planning in order to bridge the gap between information experts and responsible policy agencies (Mayer and Greenwood 1980).

Our current information society needs rapid access to statistical information. Obstacles in obtaining appropriate data on time, in usable form, may be caused by unavailability of basic data (financial flows data, for example) or underutilization of existing data bases. The latter calls in particular attention to data collection techniques. In the case of lack of data, it has to be emphasized that the widespread availability of computers today also means that users may process statistical data themselves, as is also reflected in the current popularity of machine-readable data files. Furthermore, public agencies should be aware of a wide variety of data finding aids, such as published catalogues, published indexes and table finding guides, and computerized cataloguing and query systems (see also Sprehe 1981). Recent developments in the field of decision support systems and artificial intelligence systems indicate that modern computer technology (especially with regard to data base management, data retrieval and data display) offer many new perspectives for adaptive information systems in a situation of socio-economic dynamics.

The introduction of new information and communication technologies may incorporate a mixed blessing. It may make previously routine activities redundant and may save labour costs (and hence increase both unemployment and economic efficiency). But there is on the other hand quite some evidence that new skills are required, which will add to the economic progress of our nations.

A related problem may emerge from a misunderstanding of the social meaning of information systems: it is often assumed that information systems have only a cost nature. In that case, the improvement in the quality of a decision has then to be traded off against the costs of an information system. This is, however, only a partial view on information systems. Information systems may also have many benefits, as these may lead to avoidance of the costs of taking wrong decisions. Thus the value of an information system cannot be judged without taking into account and evaluating the costs and benefits of all relevant courses of action (including the opportunity cost of information).

Given the new 'technological regime' where economic activities are increasingly dependent upon information availability and use of reliable and up-to-date information systems, the position of central areas vis-a-vis peripheral areas deserves also attention. It seems plausible that core areas are in an advantageous position as far as access to information networks is concerned. But since new technology is a source of both constraints and opportunities, the question that is emerging is concerned with the new prospects and threats peripheral areas are facing as to their ability to overcome the gap separating these from core areas. More precisely, the issue at stake concerns the new role attributed to peripheral areas in the context of the new international division of labour and how they are going to cope with the competitive advantages of the core areas in the development process (Giaoutzi and Stratigea 1989).

Thus it is evident that, due to technological and economic transformations, our societies tend to become information societies (Giaoutzi and Nijkamp 1988, Naisbitt 1982). Those nations or regions having access to up-to-date information via efficient and flexible communication channels tend to have a comparative advantage, whilst those lagging behind will encounter serious difficulties in being competitive at a European scale. International exchange of information and know-how through adequate transmission channels (e.g. telecommunication) is of critical importance here.

21.2 Demands on information systems

The desires of actors in private or public agencies with regard to the quality of an information system will depend on the nature of the planning problem at hand. In general, one may assume that these claims will be higher as: the frequencies of these choice situations are lower; the range of impacts is larger; the number of spillover effects (distributional effects) to other systems is larger; the number of conflicts involved is larger; the financial implications are more substantial; the time horizon of impacts is longer; the number of decision agencies or actors is larger; and the outcomes of choices to be made are more uncertain.

The previous remarks can be illustrated by Figure 21.1, which reflects the demands on information systems as a function of the above mentioned items. The envelope curve reflects the maximum demand, while the interior centre reflects a minimum demand (in case of daily routine decisions, for example).

The previous remarks point out once more the existence of severe trade-offs in policy analysis, decision support and information provision. These trade-offs imply:

(i)   the aim of a maximum accuracy of the data input (time series, disaggregate survey date, longitudinal data, etc.);
(ii)   the aim of a maximum quality of the information system (efficiency, flexibility, coherence, etc.); and

(iii)   the aim of a well-structured choice problem (coordination, conflict management, public participation, etc.).

Figure 21.1: The demands on information systems caused by the nature of choice problems

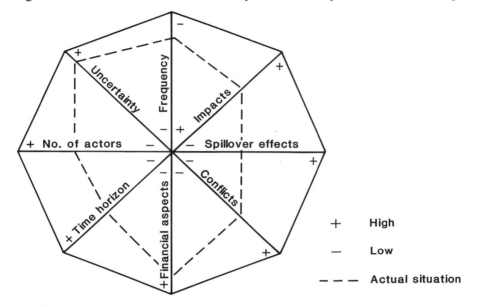

These tradeoffs can be illustrated by means of the following 'flask model', in which three flasks are connected by means of flass tubes (Figure 21.2). The three flasks are filled with water, while the three above mentioned conflictive issues (accuracy of data input, required quality of information system, and organization of choice problem), are measured on the necks of these flasks. With a given amount of effort (i.e. a given quantity of water in the flasks), it is easily seen from Figure 21.2, that a low accuracy of data will either demand a high quality of the information system or will otherwise lead to a less organized choice situation. The aim of designing a good information sysem is now to enhance the efficiency of data use and the effectiveness of policy choices, based on a well-structured transformation of data into manageable policy information (by using inter alia man-machine interactions, knowledge-based systems, connecting networks, decision support systems, and so forth).

A systematic and coherent insight into the complex pattern and evolution of a spatial system requires the design of an up-to-date, accessible and comprehensive spatial information system. Information systems for urban and regional planning should contain structured data on real-world development patterns, their properties (stability, for example), and their mutual links. Frequently, however, information systems are oriented to the national level or to specific sectors. The geographical dimension of information systems as a decision aid in urban and regional development planning has too often been neglected. Therefore, much more attention needs to be paid to the design and development of information systems reflecting socio-economic processes so as to arrive at a better representation of spatial systems and a better adaptation to the needs of urban and regional planners (Blumenthal 1969).

Figure 21.2: Tradeoffs among three items in a choice problem

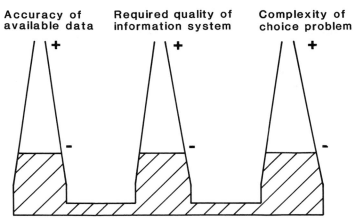

21.3 An international comparison of spatial planning and information systems

Regional and urban planning and information systems exhibit a wide variety across various countries, depending on historical, institutional and political factors. In the framework of a study carried out under the auspices of the International Institute for Applied Systems Analysis (IIASA) in Laxenburg (Austria), a cross national in-depth comparison of elements and contents of spatial information systems in six different countries has been undertaken, viz. Sweden, France, the USA, the Netherlands, Czechoslovakia (CSSR) and Finland.

Clearly, such cross-comparisons are fraught with difficulties due to the above mentioned differences among countries. The present cross-comparisons have been based on national reports on spatial information systems for the country at hand, written by an expert in this field (for more details, see Nijkamp and Rietveld 1983). All these national reports have been written according to a standard framework, so that at least a common scope and structure of the cross-national comparisons is guaranteed. Each expert from the country concerned had to provide detailed information on various planning components, linkages between these components, degree of centralisation, specific information systems for each planning component, and so forth. The following components were included in the national reports: housing, transportation, community, migration, labour market, environment and land use.

Two different steps will successively be described here, viz. a comparison of spatial planning systems and a comparison of spatial information systems. Each of these systems will be characterized by a set of attributes, while next an attempt will be made to describe the position of these systems in the six countries mentioned above by means of ordinal rankings. These rankings may not be interpreted as judgements of evaluations, but only as a way of classifying the various countries concerned. Firstly, the spatial planning systems in each of the six countries will be dealt with in a concise manner. The following attributes have been used to characterize the spatial planning systems at hand:

(i) Scope
The scope of a planning system refers to the number of planning components included in an integrated spatial planning system (transportation, housing, etc.).

(ii) Intensity
The intensity of planning activities does not only reflect the number of planning instruments, but also their potential impact (environmental controls, regional employment measures, for example).

(iii) Centralisation
Centralisation refers here to the vertical linkages between national, regional and local planning authorities. More centralisation means less flexibility and freedom for regional and local planning activities.

(iv) Coordination
Coordination should be regarded as establishing horizontal links between different planning components. Clearly, coordination may take place at each planning level (national, regional or local).

(v) Process planning
Process planning means the presence of feedback mechanisms in planning activities so as to achieve a high degree of flexibility and adjustment. Contrarily, blueprint planning refers to a situation with a clearly defined normative endpoint.

On the basis of a thorough analysis of the successive national reports on spatial planning systems, the various countries could be ranked in an ordinal sense for each of the above mentioned five attributes. The results are briefly represented in the diagram of Figure 21.3 (for more details, see Nijkamp and Rietveld 1983).

The rankings are defined in such a way that a position near the origin means a low ordinal value of the attribute concerned. The pattern reflected by Figure 21.3 is fairly remarkable. The USA have clearly a low ranking of all attributes, while the CSSR and the Netherlands have apparently an ambitious regional planning system. Intermediate positions are taken by Finland, France and Sweden.

Next, the spatial information systems in each of the above mentioned six countries will briefly be dealt with. The following attributes have been distinguished in order to characterize the successive information systems:

(i) Centralisation
A centralized information system means that a national agency is responsible for all elements of a spatial (particularly, regional) information system (including data collection, data storage and modelling).

(ii) Integration
Integration refers here to the extent to which information systems provide an integrated view of the various planning components.

(iii) Role of modelling
Modelling activities are regarded here as a way of producing new insights by means of forecasts or impact assessments.

(iv) Regional detail
Regional detail means here the level of aggregation at which information systems provide insight into the successive planning components.

(v) Computerisation

This aspect deals with the extent to which spatial information systems are based on computerisation of both inputs for and outputs of the information systems.

Figure 21.3: A cross-national comparison of spatial planning system

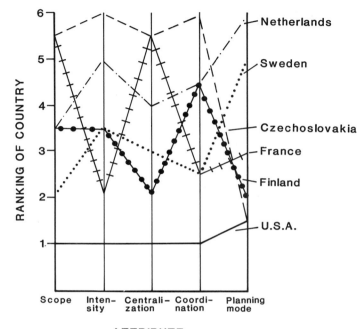

The results of the cross-national comparison of spatial information systems are summarized in the following diagram (Figure 21.4). These results exhibit much variation in information systems within each country. Fairly well developed spatial information systems appear to exist in Finland, Sweden and the Netherlands, while France and the USA appear to be marked by less developed information systems. An intermediate position is taken by the CSSR.

21.4 Geographical information systems

Introduction

Effective and accessible information systems are vital to economic performance and strategic decision-making. According to recent estimates, already more than half the jobs are directly or indirectly related to information and services, and this figure is likely to grow in the near future (Naisbitt 1984). It is increasingly believed that advanced infrastructures for information exchange and services will be as dominant in the last decade of the 20th century as canal, rail and road transport infrastructures were in previous centuries.

Especially the rapid development of digital and electronic technologies opens a new potential for sophisticated voice, data and image transmission, such as digital recording and transmission of sound and pictures, optical fibres for very fast transmission of information, super-fast computers and satellite broadcasting and video transmission. Especially in the RACE programme of the European Community many attempts are being made to stimulate and enhance R&D in information technology. Clearly, the development of hardware and software has to run parallel in this context (cf. also NCGIA, 1989).

Figure 21.4: A cross-national comparison of spatial information systems

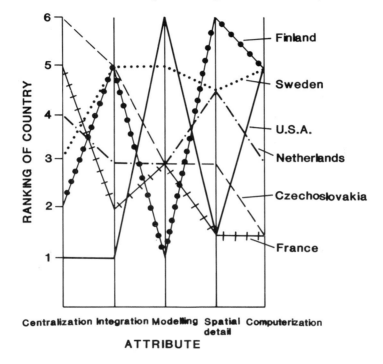

From a geographical viewpoint the trend towards advanced information systems has led to the design and use of Geographical Information Systems (GIS). A GIS serves to offer a coherent representation of a set of geographical units or objects which, besides their locational position, can be characterized by one or more attributes (feature, label or thematic component). Such information requires a consistent treatment of basic data, via the stages of collection and storage to manipulation and visualisation. In recent years, GIS has increasingly been linked to computer supported visualisation techniques such as CAD/CAM systems, as well as to remote sensing and tele-detection.

All such information systems may be highly important for planning of our scarce space, not only at a global scale (e.g. monitoring of rain forest development), but also at a local scale (e.g. physical planning). In this framework, spatial information systems are increasingly combined with pattern recognition, systems theory, topology, statistics, finite element analysis and computer-aided mapping (see also Dueker 1987). Such techniques are not only relevant for scientific research, but may also act as information bases for physical planning. The Netherlands has always been marked by a strict system of physical planning in view of a proper management of scarce space, and it is no doubt that various types of spatial information systems have been developed in recent years which serve to provide a rational basis for policy judgement.

Thus, spatial information systems have gained a great deal of importance in recent years, and planners and policy-makers are using such systems for a variety of purposes. There is a growing awareness that information is a strategic instrument with an extremely high value in a competitive or conflict-filled environment. This also explains the fact that public authorities are prepared to pay a high price for hardware and software in this area. Unfortunately, they often realize insufficiently that after the stage of buying and implementing an information system, especially high costs of maintenance and expansion, in view of new data needs, of such systems are necessary in order to render them beneficial on a structural basis for planning and policy making. In this chapter, a few observations are made regarding important aspects of GIS which are sometimes overlooked.

Product life cycle

Most GIS are still in an early stage of their product life cycle, the general pattern of which is illustrated in Figure 21.5.

Figure 21.5: A product life cycle model

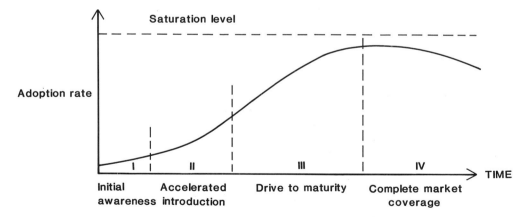

In most countries the product life cycle is now in its second phase, and it is foreseeable that further rapid developments in this area may be expected, especially if GIS is linked with statistical/econometric methods and strategic planning approaches.

Polyvalence

Information systems have often been designed for specific purposes: transportation data for transportation policy, housing data for housing market policy, and so on. Clearly, there are various rational (institutional and technical) reasons why integration of information systems is hard to achieve in the current planning practice. On the other hand, a lack of integration means an enormous waste of effort. It would already be a significant step forward if national or regional Bureaux of Statistics were authorized to provide uniform rules for data collection and standard classifications for economic activities, even if it would concern information systems beyond the responsibility of these national and regional bureaux. Experience in technical and medical sciences have demonstrated the power of uniform rules and classifications and there is no logical reason that would prevent the design of a standard frame of reference for the design of spatially-oriented information systems. This would also increase the polyvalence of such systems.

The popularity of spatial information systems is indeed to a large extent due to their polyvalence. They are able to include a variety of different information for a diversity of purposes, in principle associated with a multi-media system. This makes GIS instruments par excellence for spatial planning in a broad context. Clearly, in this initial stage of development of such systems, a lot of incoherence, overlap, redundancy and confusion is present, which is a usual feature of the introduction of a set of competitive products. Thus the polyvalent character of modern spatial information systems is a major strength, but at the same time a risk which might hamper a flexible adoption in a planning environment.

Scoping

It turns out that there is often a lack of consensus between the spatial scale of information systems and that of actual planning problems due to discrepancies between administrative (or legislative) and socio-economic regional demarcations. In this respect, a closer orientation of information systems towards administrative units is desirable. Given the multi-level pattern of regional decision making, it is in general an appropriate strategy to build up information systems in a bottom-up fashion so as to let them flexibly fit into any desired level of regional planning and policy making. Computerized information systems (geocoding, for example) may mean important progress in user-friendly data processing activities. But is is at the same clear that many efforts have to be made before complex geographical processes (filtering processes, for example) can be fully described.

The appeal of modern software for spatial information systems is its wide coverage of many attributes of objects at all possible geographical scales. However, what is often lacking is a scoping of all available information towards the specific interest of a policy-making agency. Thus customisation of the output of a spatial information system is a prerequisite in order to offer tailor-made planning instruments for geographical space. So far this has been a neglected area which needs further development in the near future. To find a compromise between over- and under-information is a basic challenge in this field.

Planning and management

In an interesting article, Goddard (1974) has pointed out the impact of corporate organizations (especially when defined in terms of the location of non-manufacturing functions and their information flow networks) on regional and urban development processes (through diffusion of technological innovation, polarized development and regional external economies, for example). The author concludes that the management of contact networks through investment in advanced telecommunications, communication audits, and the designation of growth centres based on information exchanging functions, can become an important policy instrument for both organizations and public sector agencies concerned with regional development. Thus, the contact pattern of multi-location, multi-product corporate organizations exerts an important impact on regional and urban growth processes and on the effectiveness of regional and urban policies. Communications audits administered by a public agency may become an appropriate tool in regional and urban policy by high-lighting for firms opportunities for alternative locational/organizational arrangements, especially the advantages to be gained through locating different but complementary functions in the same area (Cavill 1988).

In this context, it is noteworthy to state that in general the aim of a spatial information system is to enhance the quality of decision-making. Nevertheless, the bridge between the design and implementation of an information system on the one hand and a smooth transfer and use of such information in a planning office is difficult to build. The use of such information systems for daily management in a complex urban reality is still problematic. For instance, how to use a utility information system in a city in order to cope with all the problems of peak capacity and load management in water, telephone and electricity supply systems? The transfer of spatial information systems into management analysis is a major task. What we would need is a rigorous attempt at bridging the gap between GIS to CIA, i.e. computerized information analysis for urban and regional management. Management information systems and decision support systems are only intermediate stations on the way toward a mature CIA. Once this bridge has been built, the design of an accompanying bridge, viz. desk top planning, may be considered. Without a scientifically sound CIA desk top planning will only become a hobbyist activity of a few computer fanatics.

21.5 Conclusion

In conclusion, the modern computer hardware and software facilities provide a great deal of possibilities for coupling different information systems. In this respect, public planning agencies may learn extremely meaningful lessons from multi-plant and multi-region corporate organizations, which in general, have been able to solve the organizational problems of dealing with enormous diversified databases. In this regard, the organizational aspects of internal communication within public agencies deserve much more attention. Modern (electronic) networks provide a huge potential for improving the acquisition, communication, accessibility, efficiency, monitoring and public participation in complex public choice problems.

References

Blumenthal, S.C. (1969) Management Information Systems, Prentice-Hall, Englewood Cliffs

Burch, J.G., Strater, F.R. and Grudnitski, G. (1979) Information Systems : Theory and Practice, New York, John Wiley

Cavill, M. (1988) Handling geographic information: a telecommunication service provider's perspective, Paper 26th IGU Congress, Sydney (mimeographed)

Cordell, A.J. (1985) The Uneasy Eighties, Science Council of Canada, Ottawa

Dueker, K. (1987) Geographic Information Systems and Computer-Aided Mapping, Journal of the American Planning Association, 53 (3), 383-390

Giaoutzi, M. and Nijkamp, P. (eds) (1988) Informatics and Regional Development, Gower, Aldershot

Giaoutzi, M. and Stratigea, A. (1989) The impact of new information technologies on spatial inequalities, Paper presented at the European Regional Science Conference, Cambridge, U.K. (mimeographed)

Goddard, J.B. (1974) Organizational information flows and the urban system, Issues in the Management of Urban Systems, in Swain, H. and MacKinnon, R.D. (eds.) IIASA, Laxenburg, pp. 180-225

Mayer, R.R. and Greenwood, E. (1980) The Design of Social Policy Research, Prentice Hall, Englewood Cliffs

Naisbitt, J. (1982) Megatrends, Warner Books, New York

NCGIA (National Center for Geographic Information and Analysis) (1989) The research plan of the National Center for Geographic Information and Analysis, International Journal for Geographical Information Systems, 3 (2), 117-136

Nijkamp, P. (1988) The use of information systems for regional planning, Revue d'Economie Regionale et Urbaine, 15 (5), 759-781

Nijkamp, P. and Rietveld, P. (eds) (1983) Information Systems for Integrated Regional Planning, North-Holland Publ. Co., Amsterdam

Sprehe, J.T. (1982) A Federal policy for improving data access and user services, Statistical Reporter, March, 323-341

Wilbanks, Th.J. and Lee, R. (1986) Policy analysis in theory and practice, the regional consequences of large scale energy development, in Johansson, B. and Lakshmanan, T.R (eds), North-Holland Publ. Co., Amsterdam, pp. 304-373

Peter Nijkamp
Vrije Universiteit Amsterdam
Economische Faculteit
Postbus 7161
1007 MC Amsterdam
The Netherlands

References

Arnold, F., Jackson, M.J. and Ranzinger, M. (eds) (1986) A Study on Integrated Geo-Information Systems, EARSEL Report

Australasian Advisory Committee on Land Information (1988) Status Report on Land Information Systems in Australasia 1987-1988: Canberra, Goanna Print

Auweck, F., Bachhuber, R., Riedel, A., Theurer, R. and Schaller, J. (1989) Handbuch zur Ökologischen Bilanzierung in der Flurbereingiung, Abschlußbericht, Lehrstuhl für Landschaftsökologie TU München-Wiehenstephan, (unpublished)

Bachhuber, R. and Schaller, J. (1988) Methoden zur Umweltverträglichkeitsprüfung (UVP) agrarstruktureller Veränderungen, Schriftenreihe Bayer, Landesamt für Umweltschutz, München, 84, 65-81

Barr, A. and Feigenbaum, E.A. (1982) The Handbook of Artificial Intelligence, Volume II, Pitman, London

Bell, D.E., Keeney, R.L. and Raiffa, H. (eds) (1977) Conflicting Objectives in Decisions, International Series on Applied Systems Analysis, John Wiley

Berry, B.J.L., Marble, D.F. (1968) Spatial Analysis, Prentice-Hall, New Jersey

Berry, J.K. (1986) Learning computer-assisted map analysis, Journal of Forestry, October, 39-43

Bertin, J. (1974) Graphische Semiologie, Berlin

Besag, J.E. (1986) On the statistical analysis of dirty pictures, Journal of the Royal Statistical Society, SocB 48, 192-236

Birkin, M. and Clarke, M. (1989) The generation of individual and household incomes at the small area level using SYNTHESIS, Regional Studies, forthcoming

Birkin, M. and Clarke, M. (1988) SYNTHESIS: a synthetic spatial information system. Methods and examples, Environment and Planning, A, 20, 645-71

Birkin, M., Clarke, G.P., Clarke, M. and Wilson, A.G. (1987) Geographical information systems and model-based locational analysis: ships in the night or the beginnings of a relationship?, Working Paper 498, School of Geography, University of Leeds

Blumenthal, S.C. (1969) Management Information Systems, Prentice-Hall, Englewood Cliffs

Braedt, J. (1988) Satellitenbilder als Informationsquelle für Landesplanung und Umweltschutz, Mitteilungenblatt, Deutscher Verein Vermessungswesen, Landesverein Bayern, 40, 117-131

Breheny, M.J. (1987) Information systems and policy formulation: changing roles and requirements, in Spatial Information and their Role for Urban and Regional Research and Planning, Bundeskanzleramt, Wien, pp. 25-41

Brown, R. (1989) Technical mapping, Computer Graphics World, April, 113-117

Bryson, J.M. and Roering, W.D. (1987) Applying private sector strategic planning in the public sector, Journal of the American Planning Association, 53, 23-33

Burch, J.G., Strater, F.R. and Grudnitski, G. (1979) Information Systems : Theory and Practice, New York, John Wiley

Burrough, P.A. (1986) Principles of Geographical Information Systems for Land Resources Assessment, Monographs on Soil and Resources Survey 12, Clarendon Press

Calkins, H.W. and Tomlinson, R.F. (1984) Basic Readings in Geographic Information Systems, SPAD Systems Ltd, Williamsville, New York

Cammen, H. van der (1980) De sociale wetenschappen in het ruimtelijk beleid: twee opvattingen, in Opstellen over planologie en demografie, Planologisch en Demografisch Instituut, Amsterdam, pp. 61-77

Cavill, M. (1988) Handling geographic information: a telecommunication service provider's perspective, Paper 26th IGU Congress, Sydney (mimeographed)

Chickering, A.W., Halliburton D. et al. (1977) Developing the College Curriculum, Washington

Chorley Committee (1987) Handling Geographic Information, Report to the Secretary of State for the Environment of the Committee of Enquiry into the Handling of Geographic Information chaired by Lord Chorley, Department of the Environment, London

Clarke, K.C. (1986) Recent trends in geographic information system research, Geo-Processing, 3, 1-5

Clarke, M., Duley, C. and Rees, P.H. (1989) Micro-simulation models for updating household and individual characteristics in small areas between censuses: demographics and mobility, Paper presented to the International Migration Seminar, Gävle, Sweden, (January)

Cochrane, J.L. and Zeleny, M. (eds) (1973) Multiple Criteria Decision Making, University of South Carolina Press, Columbia

Codd, E. (1970) A relational model of data for large shared data banks, Communications of the ACM 13, 377-387

Cordell A. (1985) The Uneasy Eighties; The Transition to an Information Society, Science Council of Canada, Background Study 53, Hull

Cordell, A.J. (1985) The Uneasy Eighties, Science Council of Canada, Ottawa

Crama, Y. and Hansen, P. (1983) An introduction to the ELECTRE research programme, In P. Hansen (ed) Essays and Surveys on Multiple Criteria Decision Making, Springer, Berlin, pp. 31-42

Creusen, M. and Mantelaers, P. (1987) Teaching the design of information systems: an educational problem, Informatie, 5, 429-438, (original in Dutch)

Crosswell, P.L. and Clark, S.R. (1988) Trends in automated mapping and geographic information system hardware, Photogrammetric Engineering and Remote Sensing, 54(11), 1571-1576

Curran, P.J. (1985) Principles of Remote Sensing, Longman

Dangermond, J. (1983) Selecting new town sites in the United States using regional databases, in Teicholz, E. and Berry, B.J.L. (eds) Computer Graphics and Environmental Planning, Prentice Hall Inc., Englewood Cliffs

Dangermond, J. (1988) GIS trends and comments, ARC News, 13-17

Dataquest Inc. (1989) GIS: the new business opportunity, Business Week, 4/4/89

Date, C. (1981) An Introduction to Database Systems, Addison-Wesley, Massachussets

Department of Environment (1987) Handling Geographic Information, HMSO, London

Doukidis, G.I., Land, F. and Miller, G. (1989) Knowledge Based Management Support Systems, Ellis Horwood Ltd, Chichester

Douw, L., van der Giessen, L.B., and Post, J.H. (1987) De Nederlandse landbouw na 2000; Een verkenning, Mededeling 379, Landbouw Economisch Instituut, Den Haag

Drummond, J. (1988) Fuzzy sub-set theory applied to environmental planning in GIS, Proceedings of the Eurocarto Seven Conference on Environmental Applications of Digital Mapping, Enschede, (September 20-22)

Dueker, K. (1987) Geographic Information Systems and Computer-Aided Mapping, Journal of the American Planning Association, 53 (3), 383-390

ESRI (1988) Lärmgutachten zu den neuen Flugrouten Flughafen München II, Auszug: Lärmzonenberechnung nach AzB, Erstellung im Auftrag der Gemeinde Kranzberg,

ESRI (1989) Marketing information for ARC/INFO, ESRI, California

Faludi, A. (1973) Planning Theory, Pergamon Press, Oxford

Fedra, K. (1985) Advanced Decision-oriented Software for the Management of Hazardous Substances. Part I: Structure and Design, CP-85-18; Part II: A Prototype Demonstration System, CP-86-10, International Institute for Applied Systems Analysis, Laxenburg, Austria

Fedra, K. and Diersch, H-J. (1989) Interactive groundwater modeling: color graphics, ICAD and AI, in Proceedings of the International Symposium on Groundwater Management: Quantity and Quality, Benidorm, Spain (2-5 October)

Fedra, K. and Loucks, D.P. (1985) Interactive computer technology for planning and policy modeling, Water Resources Research, 21(2), 114-122

Fedra, K. and Otway, H. (1986) Advanced Decision-oriented Software for the Management of Hazardous Substances. Part III: Decision Support and Expert Systems: Uses and Users, CP-86-14, International Institute for Applied Systems Analysis, Laxenburg, Austria

Fedra, K. and Reitsma, R. (1989) Decision support and GIS, Paper prepared for the GIS Summer Institute, Amsterdam, (August 14-25)

Fedra, K., Karhu, M., Rys, T., Skoc, M., Zebrowski, M. and Ziembla, W. (1987) Model-based Decision Support for Industry--Environment Interactions. A Pesticide Industry Example, WP-87-97, International Institute for Applied Systems Analysis, Laxenburg, Austria

Fedra, K., Li, Z., Wang, Z. and Zhao, C. (1987) Expert Systems for Integrated Development: A Case Study of Shanxi Province, The People's Republic of China, SR-87-1, International Institute for Applied Systems Analysis, Laxenburg, Austria

Fedra, K., Weigkricht, E., Winkelbauer, L. (1987) A Hybrid Approach to Information and Decision Support Systems: Hazardous Substances and Industrial Risk Management, RR-87-12, International Institute for Applied Systems, Laxenburg, Austria (Reprinted from Economics and Artificial Intelligence, Pergamon Books Ltd)

Fedra, K., Zhenxi Li, Zhuongtuo Wang and Chun Zhao (1987) Expert Systems for Integrated Development, A Case Study of Shanxi Province, The People's Republic of China, IIASA Report, Laxenburg

Fick, G. and Sprague, R.H., Jr. (eds) (1980) Decision Support Systems: Issues and Challenges, Proceedings of an International Task Force Meeting (June 23-25), IIASA Proceedings Series, Pergamon Press, Oxford

Frank, A.V. (1988) Requirements for a database management system for a GIS, Photogrammetric Engineering and Remote Sensing, 54(11), 1557-1564

Furst, J., Nachtnebel, H.P. and Remmel, I. (1989) Ökologische Rahmenuntersuchung Straubing-Vilshofen/Grundwassermodell, Vorbericht Variantenstudie, Universität für Bodenkultur, Wien, (unpublished)

Geertman, S.C.M. (1989) The application of a geographic information system in a policy environment: allocating more than one million dwellings in the Randstad Holland in the period 1990-2015, ARC/INFO Fourth Annual ESRI European User Conference, Rome, Italy

Geertman, S.C.M. and Toppen, F.J. (1989) Voorstel voor het Hoofdonderzoek Ruimte voor de Randstad, Ruimte voor de Randstad, deel 2, Faculty of Geographical Sciences, University of Utrecht, Utrecht

Giaoutzi, M. and Nijkamp, P. (eds) (1988) Informatics and Regional Development, Gower, Aldershot

Giaoutzi, M. and Stratigea, A. (1989) The impact of new information technologies on spatial inequalities, Paper presented at the European Regional Science Conference, Cambridge, U.K. (mimeographed)

Goddard, J.B. (1974) Organizational information flows and the urban system, Issues in the Management of Urban Systems, in Swain, H. and MacKinnon, R.D. (eds.) IIASA, Laxenburg, pp. 180-225

Groot, R. (1987) Geomatics: a key to country development?, Paper of the Eleventh United Nations Regional Cartography Conference for Asia and the Pacific, Bangkok, (5-16 January)

Haber, W. and Schaller, J. (1988) Connectivity in Landscape Ecology, Proceedings of the 2nd International Seminar of the 'International Association for Landscape Ecology', Hrsg.: K.F. Schreiber, Münstersche Geographsiche Arbeiten, 29, 181-190

Harris, B. (1989) Integrating a land use modelling capability within a GIS framework, mimeo (copy available from the author)

Harts, J.J. and Ottens, H.F.L. (1987) Geographic Information Systems in the Netherlands: application, research and development, in Proceedings of the International Geographic Information Systems (IGIS) Symposium: The Research Agenda, Volume III (Applications and Implementation), Arlington, Virginia, USA, pp. 445-457 (November 15-18)

Heijden, R.E.C.M. van der (1986) A decision support system for the planning of retail facilities, Technische Universiteit, Eindhoven

Herwijnen, M. van and Janssen, R. (1988) DEFINITE, A support system for decisions on a finite set of alternatives, Proceedings of the VIIIth International Conference on MCDM, Manchester

Herwijnen, M. van, Janssen, R. and Rietveld, P. (1989) Een multicriteria analyse van alternatieve aanwendingen van landbouwgrond, Institute for Environmental Studies, Amsterdam

Hobbs, B.F. (1984) Regional energy facility models for power system planning and policy analysis, in B. Lev et al., Analytic Techniques for Energy Planning, Elsevier Science Publishers, B.V. Amsterdam, pp. 53-66

Huzen, L. and van der Schuit, J.H.R. (1989) An overview of environment related information systems at the National Physical Planning Agency, Paper presented at the Seminar on Environmental Mapping, ITC, Enschede, (April 25-27)

International Journal of Geographic Information Systems (1989) Volume 3, Number 2, 129

ITC (1987a) Draft Strategic Plan, ITC, Enschede

ITC (1987b) Provisional Syllabus Special Course for the Modern Cartographic Center of India, ITC, Enschede

Jackson, M. and Mason, D. (1986) The development of integrated Geo-Information Systems, International Journal of Remote Sensing 7 (6), 723-740

Janssen, R. (1989) Beslissings Ondersteunend Systeem voor Discrete Alternatieven, Beschrijving en handleiding, IVM, Amsterdam

Janssen, R. and Rietveld, P. (1985) Multicriteria evaluation of land-reallotment plans: a case study, Environment and Planning A, 17, 1653-1668

Jong, T. de and Ritsema van Eck, J. (1989) GIS as a tool for human geographers: recent developments in the Netherlands, Paper presented at the Deltamap User Conference, Fort Collins, USA, (April 11)

Jong, W.M. de (1989a) Uncertainties in data quality and the use of GIS for planning purposes, in Proceedings of the UDM Symposium, 1989, Lisboa, pp. 171-186

Jong, W.M. de (1989b) Development of a large national geographic database in the Netherlands, Paper presented to the 1989 SORSA Colloquium, Maryland, USA, (March 29-31)

Jong, W.M. de (1989) GIS database design: experiences of the Dutch National Physical Planning Agency, Chapter 5 of this book

Keeney, R.L., and Raiffa, H. (1976) Decisions with Multiple Objectives, Wiley, New York

Keeney, R.L. and Raiffa, H. (1976) Decisions with Multiple Objectives: Preferences and Values Tradeoffs, Wiley, New York

Khadija Haq (ed) (1985) The informatics revolution and the developing countries, Paper prepared for the North South Roundtable Consultative Meeting in Scheveningen, The Netherlands, (Sept. 13-15)

Killen, J.E. (1983) Mathematical Programming for Geographers and Planners, Croom Helm, London

Klir, G.J. and Folger, T.A. (1988) Fuzzy Sets, Uncertainty, and Information, Englewood Cliffs, New Jersey

Knoers, F. and Van Der Hoogen, J. (1981) How to design a curriculum, University of Nijmegen, Nijmegen

Koeppel, J.G., Mayer, F., Schaller, J. and Steib, W. (1988) Konzept der Ökologischen Rahmenuntersuchung zum geplanten Donauausbau zwischen Straubing und Vilshofen (BRD), Wissenschaftl, Kurzreferat zur Arbeitstagung der IAD in Mamaia/Rumänien, (in preparation)

Kreukels, A.M.J. (1985) Planning als spiegel van de westerse samenleving: de frontline van de nieuwe grootstedelijke plannen in de Verenigde Staten in de jaren tachtig, Beleid en Maatschappij, 12, 311-324

Lang, L. (1988) GIS technology puts ESRI on the map, Computer Graphics World, February, 91-94

Le Clercq, F. (1990) Information supply to strategic planning, Environment and Planning B, Pion Ltd

Levine, R. (1988) Visualization barriers, Computer Graphics World, August, 28-35

Linden G. (1987) The development of curricula in the field of Geo-Information Systems: the ITC experience, Paper presented at the 2nd International Seminar on Information Systems for Government and Business, UNCRD, Kawasaki

Lorie, R. and Meier, A. (1984) Using a relational DBMS for geographical databases, GeoProcessing, 2, 243-257

Manegold, J. (1989) NETNET, Programm zur Trittsteinbewertung, ESRI, Kranzberg, unveröffentlicht

Manheim, M.L. (1989) Towards true executive support: managerial and theoretical perspectives, in Widemeyer G. (ed) DSS 89 - Proceedings of the Ninth International Conference on Decision Support Systems, The Institute of Management Sciences, Providence, R.I.

Marble, D.F. and Peuquet, D.J. (1983) Geographic information systems and remote sensing, Manual of Remote sensing, 2nd Edition, American Society for Photogrammetry and Remote Sensing, Falls Church, Virginia

Martin, D. (1986) Advanced Database Techniques, The MIT Press, Cambridge, Massachussets

Mayer, R.R. and Greenwood, E. (1980) The Design of Social Policy Research, Prentice Hall, Englewood Cliffs

McLaughlin, J. (1988) Geographical Information System Concepts, Proceedings of the GIS Seminar in Toronto: Queen's Printer of Toronto, pp. 10-25

Meulen, G.G. van der (ed) (1988) Informatica en Ondersteuning van Ruimtelijke Besluitvorming, Eindhoven

Ministerie van Volkshuisvesting en Ruimtelijke Ordening (1988), Vierde Nota over de Ruimtelijke Ordening, deel a: beleidsvoornemen, Tweede Kamer, vergaderjaar 1987-1988, 20.490, (1 en 2), SDU, 's-Gravenhage

Ministry of Housing, Physical Planning and Environment (1988) On the Road to 2015, Comprehensive summary of the Fourth Report on Physical Planning in the Netherlands, SDU Publishers, The Hague

Naisbitt, J. (1982) Megatrends, Warner Books, New York

NCGIA (National Center for Geographic Information and Analysis) (1988) Education Plan, University of California, Santa Barbara

NCGIA (National Center for Geographic Information and Analysis) (1989) Research Initiatives, NCGIA Update, 1

NCGIA (National Center for Geographic Information and Analysis) (1989) The research plan of the National Center for Geographic Information and Analysis, International Journal for Geographical Information Systems, 3 (2), 117-136

Newell, A and Simon, H.A. (1972) Human Problem Solving, Prentice-Hall, Englewood Cliffs

Nijkamp, P. (1988) The use of information systems for regional planning, Revue d'Economie Regionale et Urbaine, 15 (5), 759-781

Nijkamp, P. and Jong, W.M. de (1987) Training needs in information systems, Development Dialogue, 8(1), Spring

Nijkamp, P. and Jong, W.M. de (1988) Informatica en ruimtelijk beleid, in Meulen, G.G. van der (ed) Informatica en Ondersteuning van Ruimtelijke Besluitvorming, Eindhoven

Nijkamp, P. and Rietveld, P. (eds) (1983) Information Systems for Integrated Regional Planning, North-Holland Publ. Co., Amsterdam

Nijkamp, P. and Rietveld, P. (1986) Multiple objective decision analysis in regional economics. In P. Nijkamp (ed) Handbook of Regional Economics, North Holland Publishing Co., Amsterdam

Nijkamp, P., Rietveld, P. and Voogd, H. (1989) Multicriteria Evaluation in Physical Planning, North Holland, Amsterdam

O'Brian, W.R. (1985) Developing "Expert Systems": contributions from decision support systems and judgement analysis techniques, R&D Management, 15(4), 293-303

Openshaw, S. (1984) The modifiable areal unit problem, CATMOG 38, Geo Abstracts, Norwich

Openshaw, S. (1988) Building an automated modelling system to explore a universe of spatial interaction models, Geographical Analysis, 20, 31-46

Openshaw, S. (1989a) Towards a spatial analysis research strategy for the Regional Research Laboratory initiative, NE.RRL Research Report 1, CURDS, University of Newcastle-upon-Tyne

Openshaw, S. (1989b) Recent developments in Geographical Analysis Machines, NE.RRL Research Report 2, CURDS, University of Newcastle-upon-Tyne

Openshaw, S. (1989c) Making geodemographics more sophisticated, Journal of the Market Research Society, 31, 111-131

Openshaw, S. and Mounsey, H. (1987) Geographic information systems and the BBC's Domesday interactive videodisk, International Journal of Geographical Information Systems, 1, 173-179

Openshaw, S., Charlton, M., and Cross, A.E. (1989) A geographical correlates exploration machine, NE.RRL Research Report 3, CURDS, University of Newcastle upon Tyne

Openshaw, S., Charlton, M., and Wymer C. (1987) A Mark 1 Geographical Analysis Machine for the automated analysis of point data, International Journal of GIS, 1, 335-343

Openshaw, S., Charlton, M., Craft, A.W., and Birch, J.M. (1988) An investigation of leukaemia clusters by use of a geographical analysis machine, The Lancet, February 6th, 272-273

Padding, P. (1987) De toekomst van de landbouw in ruimtelijk perspectief, een toepassing van het GIS ARCINFO, Geografisch Instituut, Rijksuniversiteit Utrecht and Rijksplanologische Dienst, Den Haag

Parker, H.D. (1988) The unique qualities of a geographic information system: a commentary, Photogrammetric Engineering and Remote Sensing, 54(11), 1547-1549

Pearl, J., Leal, A. and Saleh, J. (1982) GODDESS: A Goal Directed Decision Supporting Structuring System, IEEE Trans. Pattern Analysis and Machine Intelligence, PAMI, 4(3), 250-262

Pellenbarg, P., Schuurmans, F. and Wouters, J. (1974) De Ontwikkelmogelijkheden van Medemblik: Proeve van een Feasibility Study. 3. The development possibilities of Medemblik: a feasibility study, Geografische Instituut, Rijks Universiteit Groningen, The Netherlands

Phillips, L. (1984) Decision support for managers, in H. Otway and M. Peltu (eds) The Managerial Challenge of New Office Technology, Butterworths, London, p246

Planungsburo Dr Schaller (1988) Neue Schritte beim Donauausbau - Die ökologische Rahmenuntersuchung, In RMD-Intern 3/88 und 1/89, Rhein-Main-Donau AG (Hrsg.), München

Prastacos P. and Karjalainen, P. (1989) An information system for analyzing census data on microcomputers, in Ekistics (forthcoming)

Profijt, I.R. and Bakermans, M.M.G.J. (1988) Cultuurhistorische Kartering Nederland, in Geografisch Informatiesysteem, Report No. 1954, Stiboka, Wageningen

Radford, K.J. (1978) Information Systems for Strategic Decisions, Reston Publishing Co. Inc., VA

Rawstron, E.M. (1958) Three principles of industrial location, Transactions and Papers of the IBG, 25, 132-142

Reitsma, R. and Makare, B. (1989) Integration of model-based decision support and dedicated Geographical Information Systems, Journal of Geographical Information Systems (in preparation)

Reitsma, R.F. (1988) REPLACE: The Application of Relational Methodology in Site Suitability Analysis in the IIASA-ACA Shanxi Province Decision Support System, Paper presented at the Annual Conference of the Institute of British Geographers, Loughborough, UK, (January 5-8)

Rhind, D. (1986) Remote sensing, digital mapping and geographical information systems: the creation of national policy in the United Kingdom, Environment and Planning C: Government and Policy, 4, 91-102

Rhind, D. (1987) Recent developments in geographical information systems in the U.K., International Journal of Geographical Systems, 1, 229-242

Rhind, D., Openshaw, S., and Green, N. (1989) The analysis of geographical data: data rich, technology adequate, Proceedings of the IV International Working Conference on Statistical and Scientific Data Base Management, Rome, Italy, Lecture notes in Computer Science, 339, 425-454, Springer-Verlag

Rietveld, P. (1980) Multiple Objective Decision Methods in Regional Planning, North Holland, Amsterdam

Roy, B. (1968) Classement et choix en présence de points de vue multiples, RIRO, 2, 57-75

Sage, A.P. and White, C.C. (1984) ARIADNE: a knowledge-based interactive system for planning and decision support, IEEE Transactions on Systems, Man, and Cybernetics, 14(1), 35-47

Sauberer, M. (1987) Some requirements on spatial information systems at the national level, with emphasis to new planning strategies: the situation in Austria, in Spatial Information Systems and their Role for Urban and Regional Research and Planning, Bundeskanzleramt, Wien, pp. 13-16

Schaller, J. (1987) Environmental Impact Assessment (EIA) of different routing alternatives of a planned autobahn between Munich and Passau (Southern Bavaria), In the Proceedings ARC/INFO User Conference 1987, ESRI Redlands, California

Schaller, J. (1989) Environmental Impact Assessment Study for the planned Rhine-Main-Danube River Channel (Federal Republic of Germany), in Proceedings ARC/INFO User Conference 1989, ESRI Redlands, California

Schans, R. van der (1988) De geografische gegevensstroom in kaart, Geodesia, 88(4),552-554

Schilling, H. (1968) Standortfaktoren fur die Industrieansiedlung; ein Katalog fur die Regionale und Kommunale Entwicklungspolitik sowie die Standortwahl von Unternehmungen (Siting factors for industrial locations and locational choice by enterprises: a handbook for regional and communal policy making), Osterreichisches Institut fur Raumplanung, Veroffentlichungen Nr.27, Kohlhammer GmbH, Stuttgart

Scholten, H.J. and Meijer, E. (1988) From GIS to RIA: a user-friendly microcomputer-orientated regional information system for bridging the gap between researcher and user, in: Polydorides, N., URSA-NET Proceedings 1988, Athens

Scholten, H.J. and Padding, P. (1990) Working with GIS in a policy environment, Environment and Planning, B (forthcoming)

Scholten, H.J. and Padding, P. (1988) Working with GIS in a policy environment, ARC/INFO Third Annual ESRI European User Conference, Kranzberg, W.Germany

Scholten, H.J. and Vlugt, M. van der (1988) Applications of Geographical Information Systems in Europe, a state of the art, Paper prepared for the Unisys Management Seminar on GIS, Sydney, (October 25-26)

Schreiner, J. (1989) Fachliche Vorgaben zur Trittsteinbewertung, In Zwischenbericht der Projektleitung, In Planungsbüro Dr. Schaller, Arbeitsgruppe Ornithologische Arbeitsgemeinschaft Ostbayern, Projektbericht 'Ökologische Rahmenuntersuchung zum geplanten Donauausbau zwischen Straubing und Vilshofen - Bewertungsprogramm', (unpublished)

Schuit, J.H.R. van der (1989) Cartographic aspects of a central geographic database, Paper presented at the ICA Conference, Budapest, Hungary, (16-25 August)

Shoor, R. (1988) Beyond GIS, Computer Graphics World, February, 87-92

Simon, H.A. (1967) Administrative Behavior, (3rd edition), The Free Press, New York

Sprague, R.H. and Carlson, E.D. (1982) Building Effective Decision Support Systems, Prentice Hall, Englewood Cliffs

Sprehe, J.T. (1982) A Federal policy for improving data access and user services, Statistical Reporter, March, 323-341

Stephenson G. (1989) Knowledge browsing: front ends to statistical databases. In Rafanelli,M., Klensin, J. and Svensson, P., Proceedings of the 4th International Working Conference on Statistical and Scientific Database Management, Springer Verlag, Berlin

Steuer, R.E. (1986) Multiple Criteria Optimization: Theory, Computation and Application, Wiley, New York

Swaan Arrons, H. de and Lith, P. van (1984) Expert Systemen, Den Haag

Technica (1984) The SAFETI Package. Computer-based System for Risk Analysis of Process Plant, Vol.I-IV and Appendices I-IV, Technica Ltd., Tavistock Sq., London

Teichholz, E. (1988) GIS technology matures, Computer Graphics World, December, 56-60

Toppen, F.J. and Geertman, S.C.M. (1988) Ruimte voor een Miljoen Woningen in de Randstad?, Ruimte voor de Randstad, deel 1, Faculty of Geographical Sciences, University of Utrecht, Utrecht

Townsend, A., Blakemore, M., Nelson, R. and Dodds, P. (1987) The NOMIS database: availability and uses for geographers, Area, 19, 43-50

Townsend, A., Blakemore, M., Nelson, R. and Dodds, P. (1986) The National On-line Manpower Information System (NOMIS), Employment Gazette, 94, 60-64

Tufte, E.R. (1985) The Visual Display of Quantitative Information, Graphics Press, Cheshire, Connecticut

Velden, H. ten and Scholten, H.J. (1988) De betekenis van decision support systems en expert systems in de ruimtelijke planning, in Meulen, G.G. (ed) Informatica en Ondersteuning van Ruimtelijke Besluitvorming, Eindhoven

Velden, H.E. ten and Lingen, M. van (1989) Visualization and GIS, Chapter 20 of this book

Vlugt, M. van der (1989) The use of a GIS based DSS in physical planning, Paper prepared for the GIS/LIS Conference, Orlando, (November)

Voogd, H. (1983) Decision support systemen voor overheidsplanning?, Enkele introduceren-de kanttekeningen, Planologisch Memorandum 1983-4, Technische Universiteit, Delft

Voogd, H. (1983) Multicriteria Evaluation for Urban and Regional Planning, Pion, London

VROM (1989) Kiezen of Verliezen, First National Policy Plan for Environmental Protection

Wardenaar, E. (1984) Characteristics of curricula, IOWO, University of Nijmegen, Nijmegen, (original in Dutch)

Wierzbicki, A. (1983) A mathematical basis for satisficing decision making, Mathematical Modeling USA, 3, 391-405 (also appeared as RR-83-7, International Institute for Applied Systems Analysis, A-2361 Laxenburg, Austria)

Wilbanks, Th.J. and Lee, R. (1986) Policy analysis in theory and practice, the regional consequences of large scale energy development, in Johansson, B. and Lakshmanan, T.R (eds), North-Holland Publ. Co., Amsterdam, pp. 304-373

Wilson, A.G, Coelho, J., McGill, S.M. and Williams, H.C.W.L. (1981) Optimisation in Locational and Transport Analysis, John Wiley, Chichester

Wissink, G.A. (1986) Handelen en ruimte: een beschouwing over de kern van de planolo-gie, Stedebouw en Volkshuisvesting, 67, 192-194

Young, J.A.T. (1986) A U.K. Geographic Information System for environmental monitoring, resource planning & management capable of integrating & using satellite remotely sensed data, The Remote Sensing Society, Nottingham

Youngs, C.W. (1987) Policies and management of national mapping and charting programmes, Paper presented at the Eleventh United Nations Regional Cartographic Conference for Asia and the Pacific, Bangkok, (5-16 January)

Zebrowski, M., Dobrowolski, G., Rys, T., Skocz, M. and Ziembla, W. (1988) Industrial structure optimization: the PDAS model, in K. Fedra (ed) Expert Systems for Integrated Development: A Case Study of Shanxi Province, The People's Republic of China. Final Report Volume I: General System Documentation

Zimmermann, H-J. (1987) Fuzzy Sets, Decision Making and Expert Systems, Kluwer Academic Publishers

Plate 1.1   The soils map of the Netherlands: high quality GIS graphics

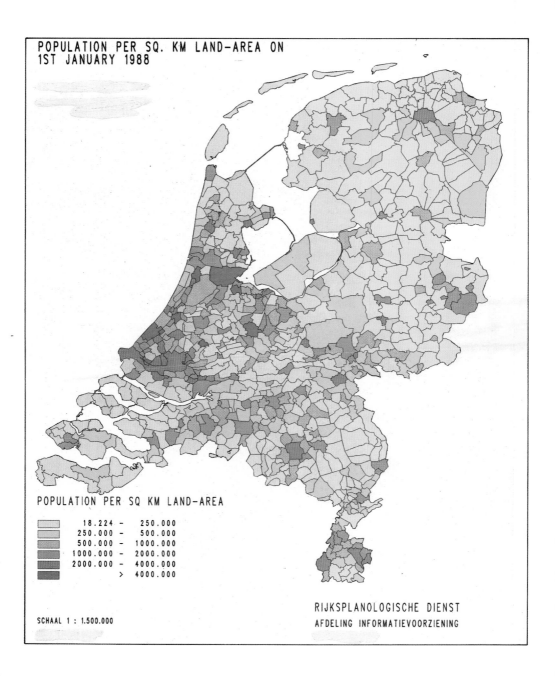

POPULATION PER SQ. KM LAND-AREA ON
1ST JANUARY 1988

POPULATION PER SQ KM LAND-AREA

18.224	–	250.000
250.000	–	500.000
500.000	–	1000.000
1000.000	–	2000.000
2000.000	–	4000.000
	>	4000.000

RIJKSPLANOLOGISCHE DIENST
AFDELING INFORMATIEVOORZIENING

SCHAAL 1 : 1.500.000

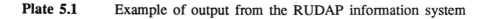

Plate 5.1   Example of output from the RUDAP information system

LANDSCAPE ECOLOGICAL MAPPING OF THE NETHERLANDS

GROUNDWATER RELATIONS

- INFILTRATION
- NO IMPORTANT INFILTRATION OR EXFILTRATION
- LOCALLY/TEMPORARILY EXFILTRATION OF 'MIX-WATER TYPE'
- EXFILTRATION OF 'MIX-WATER TYPE'
- LOCALLY/TEMPORARILY EXFILTRATION OF 'GROUNDWATER TYPE'
- EXFILTRATION OF 'GROUNDWATER TYPE'
- LOCALLY/TEMPORARILY BRACKISH EXFILTRATION
- BRACKISH EXFILTRATION
- LOCALLY/TEMPORARILY SALINE EXFILTRATION
- SALINE EXFILTRATION

This map concerns the probability of ecologically relevant
exfiltration of the indicated type, defined as a groundwater
flow towards the surface resulting in a significant effect on
the chemical and/or physical qualities of the water in the
root zone and/or the drainage system.

SCALE 1 : 1.500.000
GRIDCELL SIZE 1 KM * 1 KM
Produced by CML, SC and RPD, may 1989

Plate 5.2   Example of output from the LKN information system

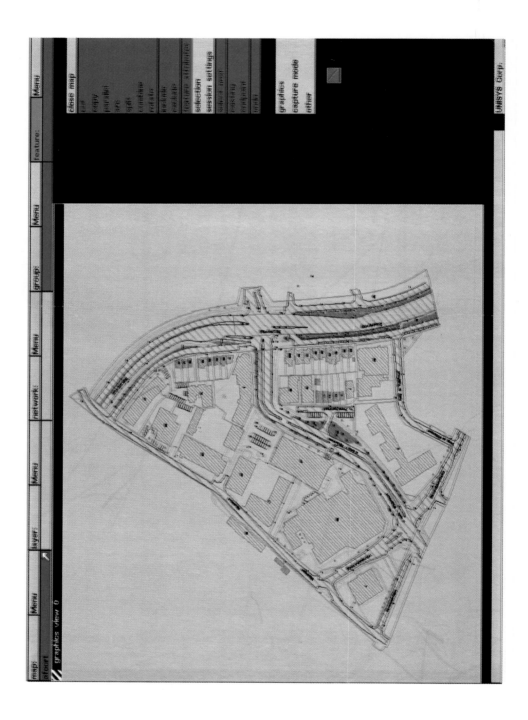

Plate 7.1   Output from ARGIS application in Amersfoort

Plate 7.2 Window facility showing attribute information

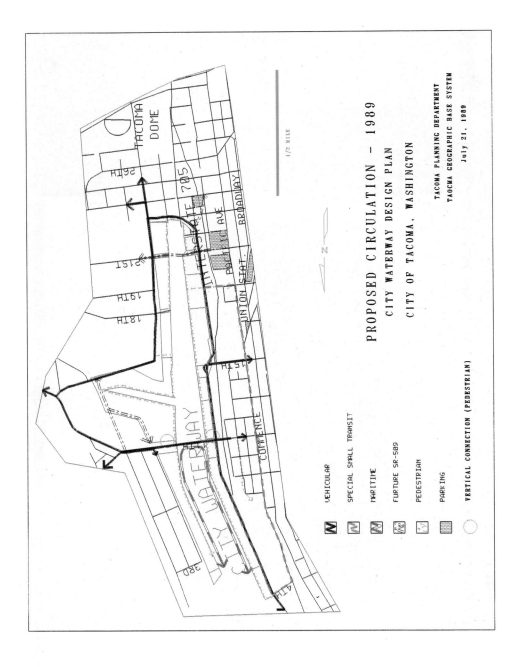

Plate 8.1   Generalised land use (parcel) map for the City Waterway
Design Plan

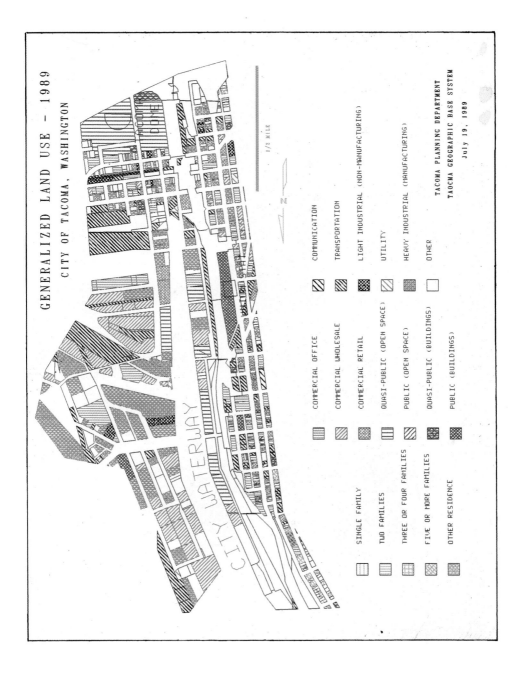

Plate 8.2   Proposed circulation (line segment) map for the City Waterway
Design Plan

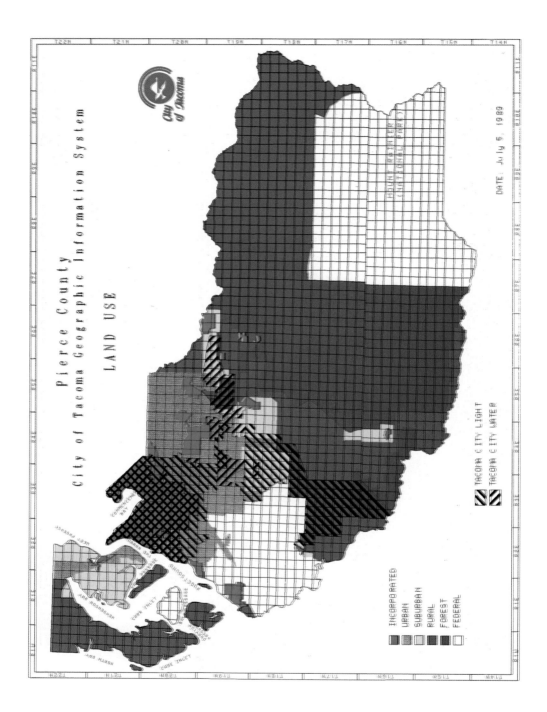

Plate 8.3   Generalised land use map of Pierce County

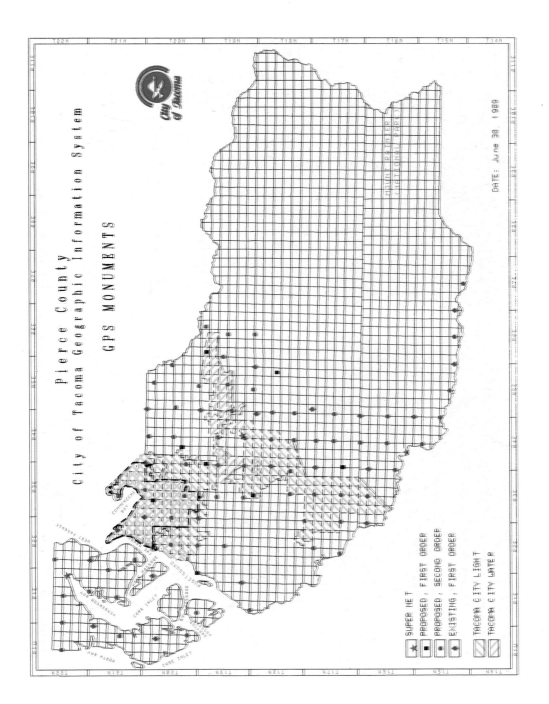

Plate 8.4   Location of GPS monuments in Pierce County

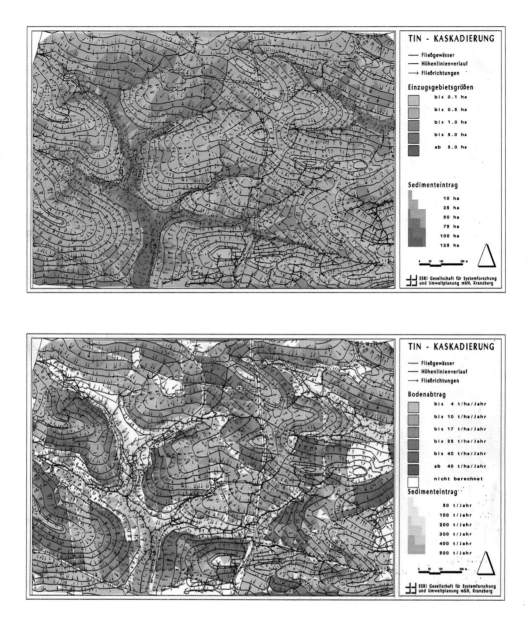

Plate 10.1   TIN cascading and calculation of soil loss

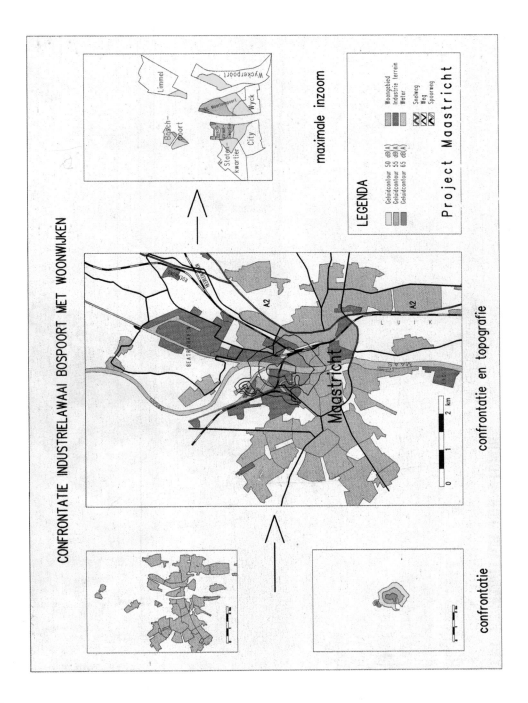

Plate 11.1   The conflict between industrial noise and residential areas in Maastricht

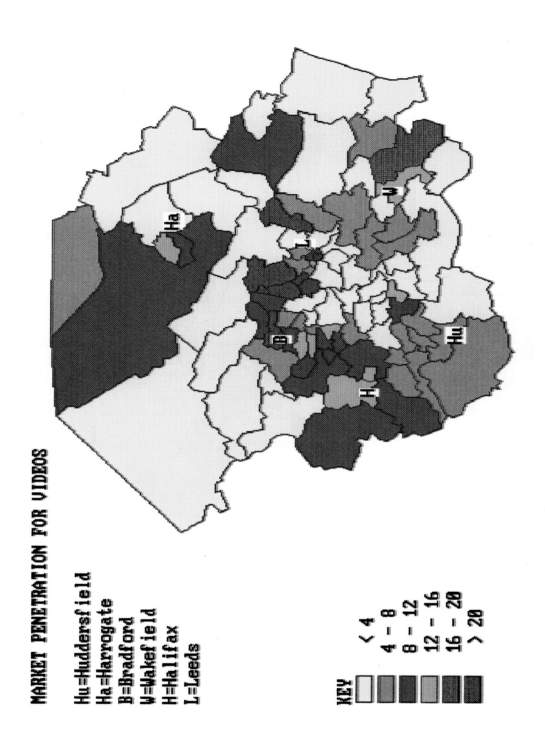

Plate 15.1   Market penetration by postal district in West Yorkshire

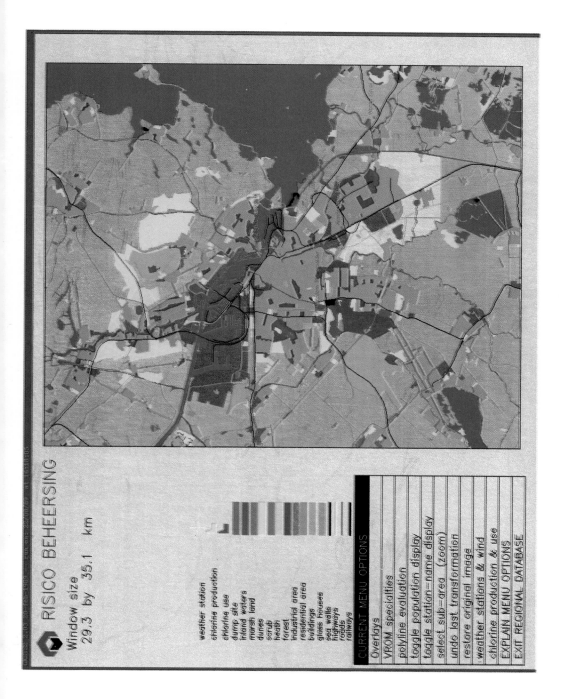

Plate 16.1   Detail from the 1:25,000 scale map of Amsterdam

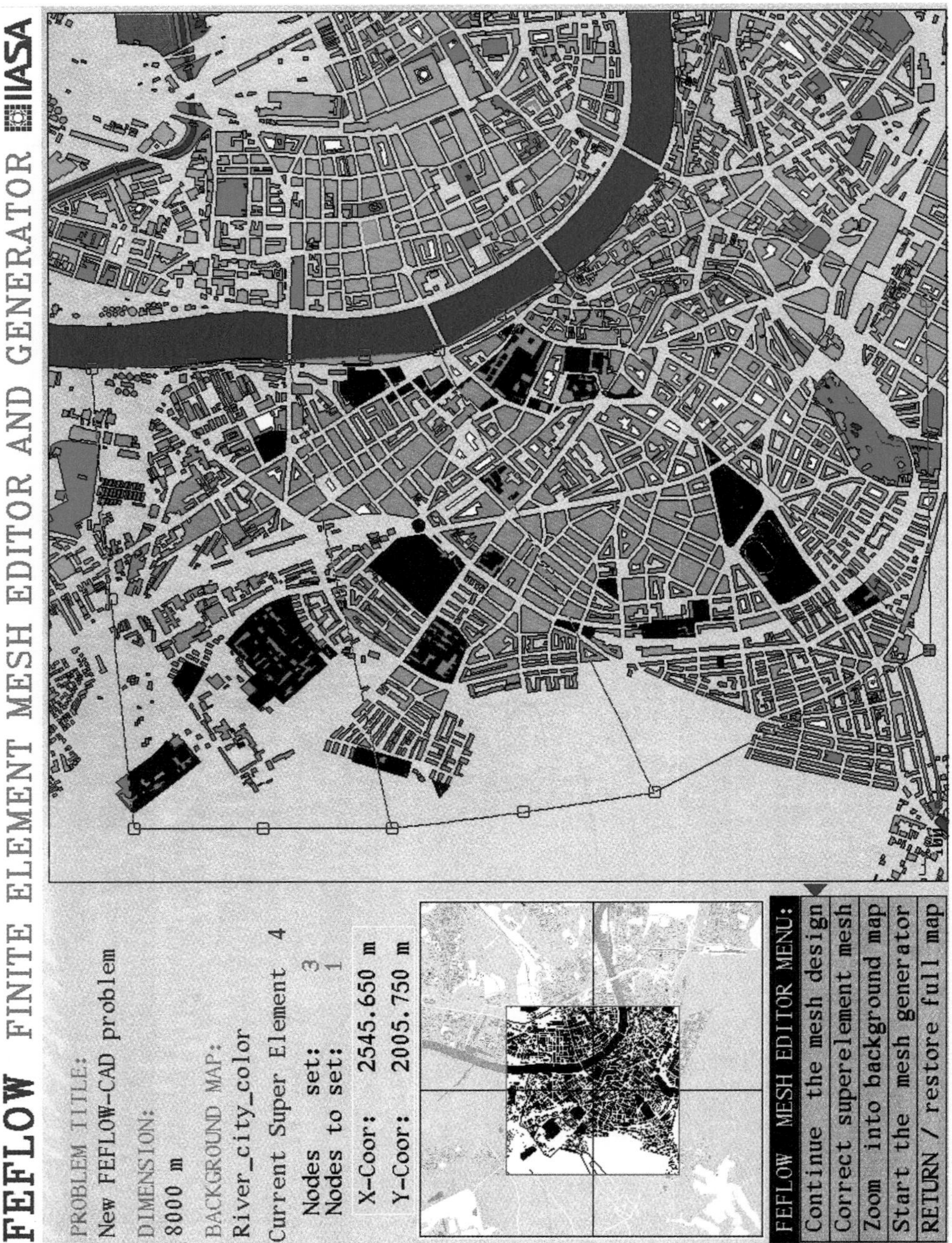

Plate 16.2 Defining a problem area in terms of super-elements